ADVANCES IN THE MICROBIOLOGY
AND BIOCHEMISTRY
OF CHEESE AND FERMENTED MILK

ADVANCES IN THE MICROBIOLOGY AND BIOCHEMISTRY OF CHEESE AND FERMENTED MILK

Edited by

F. LYNDON DAVIES and BARRY A. LAW

National Institute for Research in Dairying, Shinfield, Reading, UK

ELSEVIER APPLIED SCIENCE PUBLISHERS
LONDON and NEW YORK

ELSEVIER APPLIED SCIENCE PUBLISHERS LTD
Ripple Road, Barking, Essex, England

Sole Distributor in the USA and Canada
ELSEVIER SCIENCE PUBLISHING CO., INC.
52 Vanderbilt Avenue, New York, NY 10017, USA

British Library Cataloguing in Publication Data

Advances in the microbiology and biochemistry
of cheese and fermented milk
1. Dairy microbiology
I. Davies, F. L.   II. Law, B.A.
637'·01'576   QR121

ISBN 0-85334-287-3

WITH 25 ILLUSTRATIONS AND 28 TABLES

© ELSEVIER APPLIED SCIENCE PUBLISHERS LTD 1984

The selection and presentation of material and the opinions expressed
in this publication are the sole responsibility of the authors concerned

Printed in Northern Ireland by The Universities Press (Belfast), Ltd.

Preface

The manufacture of cultured dairy products constitutes a major fermentation industry, particularly in the developed countries. For example, in 1980, 3·5 million tonnes of cheese and over 1500 million litres of yoghurt and fermented milks were produced in the EEC. Milk and milk products account for about 15% of domestic food expenditure in the UK. Although liquid milk consumption is declining in many countries, the market for cheese is increasing slowly, with a tendency towards more complex varieties whose maturation is most difficult to control. Fermented milks enjoy not only increased interest in a wider variety of products but also an overall increase in popularity. It is clear, therefore, that the fermentation of milk to nutritious, stable and interesting products is an expanding activity which requires a constant input of both new technology and fundamental research. Although the industry can call upon a comprehensive technical literature as an aid to solving some of its problems, an equivalent coverage of basic research from which to draw new ideas is not at present available. Such information is widely scattered in specialized publications. We have therefore attempted to provide one volume in which research relating to the major stages of fermented dairy product manufacture is reviewed. In this way we hope to provide some insight into possible future directions for development in the industry, based on current efforts in research laboratories.

Studies into the physico-chemical nature of cheese have been important in investigations into modified processes such as the use of concentrated milk and the increased incorporation of whey proteins. The starter cultures used in these fermentations can now be more accurately identified and their relationships determined by the use of techniques such as DNA–DNA hybridization. The physiological mechanisms by which they utilize lactose and milk proteins to produce lactic acid and other essential end products are only now being well characterized, and the genetic control of these mechanisms is a subject of increasing study. Thus modification and improvement of starter cultures for traditional fermentations and the extension of their use to new food areas are important new possibilities. The application of genetic knowledge will also be prominent among approaches being taken to eliminate or control the attack of starter cultures by bacteriophages. Elucidation of the biochemical pathways of flavour development in fermented milks has led to new possibilities for product development and in the case of cheese this knowledge has been supplemented by important new work which could substantially reduce the traditionally long ripening periods required to achieve a mature flavour. Early detection of key flavour volatiles could be important in prediction of final mature cheese quality and the use of non-sensory instrumental methods will permit more objective assessments to be made.

It is the advances in these areas with which this book is concerned. All the contributors are actively engaged in the research they describe and the individual chapters are seen as current 'state of the art' reviews directed at those working at this level in industry, government and education.

BARRY A. LAW
F. LYNDON DAVIES

Contents

vii

Chapter 1

Milk Coagulation and the Development of Cheese Texture

Margaret L. Green

National Institute for Research in Dairying, Shinfield, Reading, UK

1. INTRODUCTION

Perhaps the most important consideration in the manufacture of cheese is to obtain an acceptable product from the point of view of both flavour and texture. The texture is the manifestation of the rheological characteristics and is dependent on the composition and structure of the cheese. Because of developments in microscopy, especially scanning electron microscopy (SEM), over the last 10 years, cheese and curd structure have become amenable to detailed study.

The cheese texture may be influenced by the processing conditions at various stages during cheesemaking. The formation and handling of the curd affect its ability to retain fat and moisture, so influencing the cheese composition and thus, the texture. The crumbliness of cheese may reflect the extent of curd fusion allowed, while the firmness may depend on the extent of proteolysis during ripening.

In this chapter, factors affecting the composition and texture of cheese will be reviewed. This will involve discussion of the mechanisms of the enzymic coagulation of milk, curd assembly and the conversion of curd to cheese, and consideration of the ways in which the characteristics of milk affect its cheesemaking potential. Methods for measuring the formation and properties of curd and the texture of cheese will also be described. Of these matters, only work on the mechanism of milk coagulation appears to have been reviewed in English in any detail recently (Dalgleish, 1982). Emphasis will be given here to work published within the last 5 years. First, however, important advances

1

relating to the use of milk coagulating enzymes, made since the author's previous review (Green, 1977), will be described.

2. MILK COAGULATING ENZYMES

The prime purpose of cheesemaking coagulants, the conversion of liquid milk to a gel, can be catalysed by many different proteases. However, part of the coagulant normally survives throughout cheese-making so that enzymic activity persists in the whey and in the final curd. That in the curd normally contributes to proteolysis during ripening. The amount and type depends on the activity and specificity of the enzyme and affects the flavour and texture developed in the cheese. If the coagulant is highly proteolytic there may also be significant hydrolysis of the casein during cheesemaking, resulting in reduced yields of curd. Residual coagulant in the whey may also cause difficulties in any further processing.

2.1. Structures and Specificities of Acid Proteases
Apart from a few coagulants used on a small scale for some local cheeses, the enzymes having significant value as cheesemaking coagulants are all acid proteases (EC 3.4.23) produced from animal stomachs or mould culture filtrates. The enzymes of this group are all endopeptidases having an acid pH optimum. Pig pepsin A (EC 3.4.23.1) is the most thoroughly studied example.

The gastric proteases are secreted as zymogens; activation requires H^+ only. At low pH, the zymogens undergo a conformational change and become enzymically active. Irreversible activation then takes place by limited proteolysis (Foltmann, 1981).

Present evidence suggests that different acid proteases, from both gastric and mould sources, have closely similar primary and secondary structures. This can cause problems in identification, since antisera against one enzyme may cross-react with others. The molecules of all the acid proteases investigated have two globular domains with a cleft between them, on either side of which is an active site aspartyl residue (Tang *et al.*, 1978). The enzymic activities are inhibited by the peptide analogue, pepstatin. Binding to immobilized derivatives of this inhibitor can form the basis of affinity chromatographic methods for isolation of enzymes of this group.

Acid proteases tend to cleave peptide bonds adjacent to hydrophobic amino acids, but secondary interactions between the enzyme and a number of amino acid residues in the substrate are important in determining specificity. The effect of size and composition on the efficiency of κ-casein analogues as substrates for chymosin (EC 3.4.23.4) has been investigated in some detail (Visser, 1981). At least a pentapeptide, Leu–Ser–Phe–Met–AlaOMe or Ser–Phe–Met–Ala–IleOMe was required for any hydrolysis to occur. There was a 250–500 fold increase in substrate efficiency with the hexapeptide, Leu–Ser–Phe–Met–Ala–IleOMe. The further addition of His–Pro–His–Pro–His– to the N-terminal end and –Pro–Pro–LysOH to the C-terminal end of the hexapeptide to make a peptide containing residues 98–111 of κ-casein caused a further 110-fold increase in substrate efficiency to approximately the level of that of κ-casein. It was concluded that residues 104–106 are specifically involved in the enzymic reaction and residues 98–103 and 108 are important in enzyme-substrate binding. Such observations provide an explanation for the view that the coagulant hydrolyses only caseins and their larger fragments during cheese ripening.

2.2. Assay and Identification of Coagulants

The assay of the enzymic activity of coagulants by milk clotting or hydrolysis of protein suffers from variability of the substrate or imprecision as to the number and identity of peptide bonds being broken, respectively. These problems can be overcome by use of a pure peptide substrate containing a single hydrolysable bond as the primary standard for assay of coagulant activity. Hydrolyses of 2 hexapeptides (Table 1) have been recommended as standard assay methods (De

TABLE 1
Specific Activities of Chymosin and Bovine Pepsin A in Hydrolysing Hexapeptide Substrates

Substrate	Specific activity (μMS^{-1}, mg^{-1})			References
	Chymosin	Bovine pepsin A	Chymosin/ bovine pepsin A	
Leu–Ser–Phe–Nle–Ala–IleOMe[a,c]	107 ± 6	890 ± 21	0·120	De Koning *et al.* (1978)
Leu–Ser–Phe(NO$_2$)–Nle–Ala–LeuOMe[b,c]	109 ± 19	2635 ± 150	0·0414	Martin *et al.* (1981)

Assay conditions: [a] 0·1 mM substrate, ionic strength = 0·05, pH 4·7, 30°C, hydrolysis detected with ninhydrin; [b] 0·2 mM substrate, ionic strength = 0·1, pH 4·7, 30°C, hydrolysis detected by ΔA_{310}.
[c] Nle; norleucine.

Koning, 1980). However, even though the assay conditions were similar, the two sets of results do not entirely agree. This may be because the standard 'pure' enzymes contained some inactive material. For a reliable standardization it would be necessary to titrate the active sites present with a labelled specific inhibitor.

In the most commonly used cheesemaking coagulant, bovine rennet, chymosin and bovine pepsin A together represent 97–100% of the total clotting activity. The definition of the coagulant requires knowledge of the content of each active enzyme. Martin *et al.* (1981) have proposed a method for determining this, based on chromatographic separation of the two enzymes and their separate assay using reconstituted milk powder, previously standardized against hexapeptide using reference enzymes.

More discriminating methods of identification are required if non-bovine enzymes may be present in a coagulant. The International Dairy Federation has recommended a procedure for qualitatively identifying the major coagulants used commercially. This involves an amylase test, polyacrylamide gel electrophoresis and isoelectric focussing (De Koning, 1980). Immunological methods are faster, simpler and can allow direct quantitation in theory, but care must be taken as antibodies may cross react and enzymically-inactive antigens may be present. Monospecific antibodies against the major commercially-available coagulating enzymes have been prepared and used successfully to identify enzymes in mixtures, even at very low levels, by double immunodiffusion (Collin *et al.*, 1982). The enzymes in mixed coagulants have been quantified directly by rocket immunoelectrophoresis using monospecific antibodies, with checks to make sure that the total coagulating activity agrees with the enzyme content (Rothe *et al.*, 1976; Harboe, 1981). Such methods would probably be used more generally if monospecific antibodies became commercially-available.

Determination of residual coagulant activity in curd or cheese requires a method which is sensitive enough to detect low levels, but selective for the coagulant enzymes. Most authors have assumed that ability to clot milk or hydrolyse κ-casein detects coagulant enzymes but not those of starter. Experience of the methods in use suggests that this view is justified, at least as a first approximation. The methods used by different authors differ in the extraction procedure used and, perhaps more importantly, in the assay method. Methods using a milk clotting assay (e.g. Stadhouders *et al.*, 1977) require extensive dialysis of the extracted enzyme followed by lyophilization. Assay by κ-casein hydrolysis is both more sensitive and less affected by other materials present in the cheese extracts (Green *et al.*, 1981a). One method has been devised which specifically detects chymosin in cheese (Matheson, 1981). Cheese extracts were applied to gels containing κ-casein, with and without overlays containing monospecific antichymosin. The concentration of chymosin was given by the difference in length of the precipitation zones in the two gels. By the use of other monospecific antibodies, it should be possible to apply the method to the quantitation of other coagulant enzymes. Alternatively, it might be simpler to develop a method combining the sensitivity of gel diffusion with the specific inhibition of acid proteases by pepstatin (Takahashi *et al.*, 1978).

2.3. Alternatives to Calf Rennet

For the last 25 years, too few calves have been slaughtered worldwide to satisfy the demand for rennet for cheesemaking. Consequently, extracts from the stomachs of more mature bovines are being offered commercially for cheesemaking use. In these coagulants, which have various designations, such as calf rennet, bovine rennet or Stabo rennet, up to 100% of the clotting activity may be due to bovine pepsin (De Koning, 1980). When compared with almost pepsin-free chymosin for commercial-scale Cheddar cheesemaking, bovine pepsin gave good quality cheese, but the yield was 0·25% lower, which is economically significant (Emmons *et al.*, 1976). On the other hand, coagulants containing 70–80% bovine pepsin caused no loss of yield or quality of either Grana (Corradini & Dieci, 1978) or Tilsit (Thomasow, 1980) cheeses. Another possibility is to use stomach extracts from other young ruminant species for cheesemaking. Studies of the preparation and properties of kid and lamb rennets suggest that these enzymes would prove suitable (Anifantakis & Green, 1980).

There is still an interest in non-ruminant coagulants, some of which are in widespread commercial use. Recently, considerable effort has been directed to assessing the potential of chicken pepsin, which is produced and used commercially in Israel for unripened and ripened cheese varieties (Gordin & Rosenthal, 1978). However, some other workers have found that the coagulant gave problems, being more proteolytic than is desirable. Stanley *et al.* (1980) confirmed earlier work in finding a crude preparation to be unsatisfactory for Cheddar cheesemaking, the cheeses having poor texture and flavour. No better results were obtained with a pure enzyme preparation (J. Kay & M. L.

Green, unpublished work). However, a mixture (chicken pepsin/calf rennet in the ratio 30/70) was used successfully for semi-hard cheeses (Hušek & Dědek, 1981). Interestingly, rabbit pepsin, which is similar to chicken pepsin in its stability at pH 7, was also found to be too proteolytic for use in cheese (Wahba & El-Abbassy, 1981).

In some countries, much of the commercial cheesemaking is carried out with microbial rennets. The coagulants have apparently been improved since their initial trials, partly by the removal of contaminants such as lipases. The application of affinity chromatography, using immobilized pepstatin derivatives, can permit fast, easy purification of several acid proteases in large quantity and high yield (Kobayashi *et al.*, 1982).

Another problem in the use of microbial enzymes as coagulants results from their high heat stability. Thus, they are not always completely destroyed by pasteurization of whey, and can cause coagulation if added to milk, or proteolysis of other materials. However, they can be destroyed by treatment at 74–80°C for 25 s provided the pH value of the whey is adjusted correctly (Thunell *et al.*, 1979). The most heat-stable of the generally-used coagulants, that from *Mucor miehei*, is now available in oxidized form. This treatment does not affect its cheesemaking properties, but renders it more labile to heat so that it is unlikely to survive normal whey processing temperatures (Phelan & O'Brien, 1982).

An alternative to using microbial enzymes as coagulants is to modify microbes to produce chymosin. To this end, the gene for producing chymosin has been incorporated into the genetic material of *E. coli* (Harris *et al.*, 1982). This raises the possibility of preparing large amounts of pure enzyme from microorganisms. This would perhaps obviate the need for other coagulants, even for 'vegetarian' cheese, although additional enzymes may be needed to effect sufficient proteolysis during ripening (see Chapter 8).

The factors affecting chymosin production in calf abomasa have also been investigated. Provided a milk diet was fed, chymosin secretion was maintained at an approximately constant level, although pepsin secretion increased steadily with age. Weaning was accompanied by a rapid decrease in chymosin secretion, although this could be reversed by feeding a milk diet again within 3 weeks (Garnot *et al.*, 1977). The same cells in the base of the gastric gland secreted both prochymosin and pepsinogen, but those in the upper part secreted prochymosin only. A relatively large number of prochymosin-secreting cells were still present in the cow, but they were almost inactive (Andrén *et al.*, 1981). If means could be found of reactivating these cells to produce prochymosin, the proportion of chymosin in rennet from weaned animals could be increased.

An alternative approach is to try to develop reusable coagulants. The method tried has been based on hydrolysis of the κ-casein in milk at less than 15°C by passage through a reactor containing insoluble coagulant prepared by linking acid proteases to carriers. If successfully treated, the milk should then coagulate on warming. Some workers have reported success with this approach (Taylor *et al.*, 1979), whereas others have attributed all the activity to soluble enzyme leached from the carrier and pointed out that steric effects would block access of the average casein micelle to the enzyme on most insoluble carriers used (Beeby, 1979; Green, 1980). Two recent reports have provided strong support for the latter view. Beeby (1979) attempted to facilitate the accessibility of the κ-casein in casein micelles to the immobilized coagulant by linking the protease to the carrier by means of long, flexible polypeptide chains. The derivatives hydrolysed whole casein satisfactorily, but showed little activity on milk, causing neither coagulation nor significant κ-casein hydrolysis. Chaplin & Green (1982) used soluble, size-fractionated dextran–pepsin conjugates, which are easier to purify and characterize than insoluble derivatives, as models. The conjugates all hydrolysed κ-casein in milk and caused coagulation, but the rates declined rapidly with size, such that no milk clotting activity was expected for derivatives larger than about 30 nm diameter. Since insoluble enzyme derivatives are of the order of 10^3 times larger, they are most unlikely to be active in milk clotting.

3. MEASUREMENT OF CHEESEMAKING PARAMETERS

Cheesemaking is concerned with the formation of a curd of the correct composition and structure and its storage under conditions where it will ripen appropriately. It requires that the loss of whey from the initial coagulum, the fusion of the curd particles and the production of lactic acid are coordinated together, so that all reach the required final values simultaneously. To achieve this end, it is often useful to isolate the individual steps in the process and investigate how these are affected by variables. However, when this is done, it is important that the results can be related back to the whole process. This requires that either the substrate is the same as used in cheesemaking, i.e. prepared from fat-containing milk at the required pH and with any necessary additions, or appropriate allowances are made for differences in substrate composition.

3.1. Curd Formation

The coagulating properties of a milk can be defined empirically by the onset of coagulation (clotting time) and the time when the curd reaches the correct firmness for cutting (cutting time), although, for cheesemaking, the rate of development of the curd may also be important. However, there are problems in practice, because different methods may give different values for the 2 times, and the measuring conditions may not always be identical with those in the cheese vat.

The clotting time can be measured most simply by visual observation. A number of instrumental methods are also available (Ernstrom, 1974) but may give different results from the visual method. For instance, clotting times measured visually, viscometrically and by damping a pendulum progressively increased (McMahon & Brown, 1982), presumably because the end points occurred at different extents of casein aggregation.

'Curd firmness' can be described as the resultant of 2 elastic and 2 viscous moduli (Scott Blair & Burnett, 1958), whereas most instruments probably detect one modulus preferentially but not exclusively. As it is rarely clear exactly what is being measured, it is hazardous to equate results obtained with different instruments or attempt to derive fundamental information. However, some of the instruments now available can be used reproducibly, enabling comparisons to be made in practical terms, and most have a continuous output, permitting both the rate of increase in firmness and the firmness at any time to be determined. Typical curves obtained with the instruments described below are shown in Fig. 1. They clearly show that instruments cannot be used interchangeably, since the curves vary in both starting point and general shape.

Vibrating-reed viscometers (Ultra Viscoson 1800, Bendix Corp. Inc., Lewisburg, W. Virginia; Unipan 505, Unipan Scientific Instruments, Warsaw, Poland) are particularly useful for detecting the early stages of milk coagulation, but can be used directly in the vat up to the cutting time (Marshall *et al.*, 1982). Most other instruments appear to measure some aspect of the rigidity and require the presence of a definite gel before giving any output. The Formagraph (Foss & Co.,

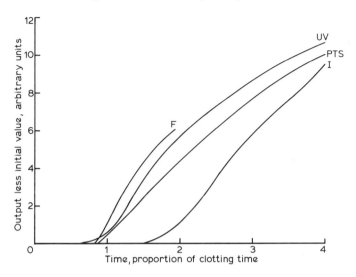

Fig. 1. Typical curves of increase in curd firmness with time. The Ultra Viscoson (UV) and pressure transmission system (PTS) were used on pasteurized whole and skim milk, respectively, as described by Marshall *et al.* (1982). The Formograph (F) was used as recommended by the manufacturers on pasteurized milk. The Instron (I) was used on raw whole milk as described by Storry & Ford (1982).

Hellerup, Denmark) derived in principle from torsion viscometers, is based on the movement of stainless steel loop pendula in 10 linearly-oscillating samples of curd. The manufacturers claim that it can provide a useful parallel to the coagulation of milk in the vat. However, the instrument has not been available long enough to allow for independent assessments, although first impressions are encouraging (Zannoni *et al.*, 1981). The Instron Universal Testing Instrument (Instron Corp., Canton, Massachusetts) has been used as a penetrometer, enabling the force exerted when the curd is compressed (Burgess, 1978) or cut (Storry & Ford, 1982) to be recorded. Vanderheiden (1976) built a hydraulically-operated instrument for measuring the rigidity of curd by its ability to transmit a pressure wave (pressure transmission system, PTS). It could be used directly in the vat. Preliminary assessment trials suggest that it can be used to indicate the optimum cutting time for Cheddar curd (Kowalchyk & Olson, 1978). An instrument based on the same principle, but electronically-operated, was developed by Hatfield (1981). It could be used satisfactorily directly in the vat, but its value for indicating the correct cutting time has not yet been assessed. It was used in parallel with a vibrating-reed viscometer in laboratory experiments, and was generally found to start responding later after rennet addition but to be more sensitive to the later stages of curd formation (Fig. 1), and to be less affected by experimental variables (Marshall *et al.*, 1982).

Only a limited amount of work has been done in which the rheological properties of curd have been described in theoretically-sound physical terms (Scott Blair, 1971; Douillard, 1973). The instruments employed in these early studies were very difficult to use accurately, but possible alternatives have become available more recently. Oscillating concentric cylinder-type viscometers have been applied successfully to measuring the shear modulus of curd (Tokita *et al.*, 1982; Van Dijk, 1982). A pulse shearometer, with which the shear modulus can be determined from the measured propagation velocity of a shear wave through the material, has been used mostly for polymer latex gels (Goodwin & Khidher, 1976), but was also found to be suitable for curd (M. L. Green, unpublished work). With such instruments, it should be possible to compare the rheological properties of curds derived from milk with those of other gels, which might help in understanding the forces involved in curd assembly.

3.2. Syneresis of Curd

Syneresis comprises loss of whey and shrinkage of the curd, either of which can be measured. In practice, most methods depend on measuring the whey volume either by separating it from the curd or from the extent of dilution of a tracer. Problems arise from the fragility of the curd and from the influence of curd manipulation and whey removal on the rate of syneresis. To overcome these, Marshall (1982) has developed a method in which the curd is held in a beaker by means of a light wire grid and the whey is poured off at fixed intervals. The method is easy to use and reproducible, and has enabled the influence of a number of variables on the syneresis rate to be investigated. The tracer dilution method requires stirring the cut curd in the whey, such as normally occurs during cheesemaking. The tracer must be a material which does not adsorb to the curd and can be added in a small volume. Blue dextran has been found to be a suitable tracer, and a reproducible method has been described (Zviedrans & Graham, 1981).

3.3. Curd Fusion

Measurement of the extent of fusion of curd particles has received little attention as yet. However, it can be determined from the frequency of holes, originally whey-filled, in electron micrographs of curd at the appropriate stage (R. J. Marshall, personal communication). A more direct method, suitable for use in pressed curd, might be developed from a photographic method for identifying curd granule junctions in Cheddar cheese (Kalab *et al.*, 1982).

3.4. Moisture Content of Curd and Cheese

The BSI standard method for determining the moisture content of curd and cheese involves drying the material at 102°C for 4 h. Microwave heating forms the basis of a much more rapid method, which takes 10–20 min in total for the drying and cooling stages (Shanley & Jameson, 1981).

4. MECHANISMS OF MILK COAGULATION, CURD ASSEMBLY AND SYNERESIS

Cheese consists of a continuous hydrated casein matrix containing discrete droplets of fat. Varieties differ in the amounts of water, salts and acid associated with the casein, the relative proportions of the phases and their structural interrelationships, and the degree and type of protein and fat breakdown permitted. The assembly of this structure is initiated by the formation of a casein network in which the fat globules are entrapped.

4.1. Enzymic Stage of Milk Coagulation

Milk coagulation occurs in two discrete steps, the enzymic initiation phase being essentially complete before aggregation starts. The initiation phase comprises the specific proteolysis of the micellar κ-casein to para-κ-casein by rennet or some other enzyme. This process has usually been followed by determining the amount of the heterogeneous macropeptide released, as either N- or sialic acid-containing material soluble in 2% or 12% (w/v)-trichloroacetic acid (TCA) (Ernstrom, 1974). However, only carbohydrate-containing macropeptides are soluble in 12%-TCA. All the macropeptides are extracted by 2%-TCA, but are difficult to determine accurately because of the large amount of background nitrogenous material which is also soluble. The method has been simplified by measurement of the fluorescence after reaction of the product with fluorescamine (Pearce, 1979) but without any fundamental effect on the disadvantages. However, some improvements in accuracy and specificity have been made by the use of a high performance liquid chromatograph to separate and measure the macropeptide from an 8% (w/v)-TCA extract (van Hooydonk & Olieman, 1982). More fundamentally, the inevitable disadvantages of using the measurement of only a proportion of a heterogeneous product to follow the reaction have been overcome by determining the homogeneous product, para-κ-casein, by quantitative polyacrylamide gel electrophoresis (Chaplin & Green, 1980).

There is strong evidence that κ-casein is located on the surface of casein micelles and that it is all hydrolysed when milk coagulates. Probably the hydrolysable portion (the hydrophilic macropeptide comprising about 30% of the molecule and containing all the phosphorus and carbohydrate) projects from the surface of the micelle into the solvent (Walstra *et al.*, 1981). These projections seem to exert a steric or entropic repulsion which keeps native casein micelles apart (Walstra, 1979). It is only when at least 85% have been removed by enzymic hydrolysis that the denuded micelles can start to aggregate (Chaplin & Green, 1980).

4.2. Aggregation Phase

The aggregation phase, which can be followed quantitatively by turbidimetry (Dalgleish, 1982) or electron microscopy (Green & Morant, 1981) occurs by a random, diffusion-controlled mechanism. Thus, the process can be described by the modified Smoluchowski equation:

$$1/n_t - 1/n_0 = 4\pi D R^* \varepsilon t \qquad (1)$$

where n_0 and n_t are the numbers of particles present initially and after time t, D is the diffusion coefficient, R^* is the collision radius and $\varepsilon = \exp(-W/RT)$, where W is the energy barrier rendering collisions elastic. (The bimolecular rate constant for aggregation, $k_s = 8\pi D R^* N_{AV} \varepsilon$, where N_{AV} is Avogadro's number.) If eqn (1) is combined with others taking account of the enzymic hydrolysis of κ-casein and the stability of incompletely-hydrolysed micelles to aggregation, the complete aggregation process can be modelled mathematically. Two such equations have been derived independently (Dalgleish, 1982; Darling & van Hooydonk, 1981), but they differ little in the assumptions on which they are based and both predict the observed effects of a number of variables on the clotting time of milk.

Dalgleish *et al.* (1981) showed the rate of aggregation of casein micelles at 21°C to be independent of their size and little affected by doubling the concentration of rennet. Thus, the rates observed by these workers were close to the maximum rate for the chosen conditions. However, the value of k_s was only 10^{-2}–10^{-3} of the diffusion-controlled limit. This can be explained if the energy barrier (W) to effective collisions is about $15 \, kJ \, mol^{-1}$.

The slower than theoretical rate of aggregation of rennet-treated micelles could reflect either a true energy barrier, i.e. a repulsive force hindering the close approach of two micelles, or a steric factor, i.e. the need for micelles to approach one another in the correct orientation for aggregation to occur. Rennet-treated casein micelles carry a net negative surface charge, but the range of this is probably limited by the shielding effect of the soluble ions in milk to about 0.1 nm (Payens, 1977). This distance is probably comparable to irregularities in the micellar surface, so it is unlikely that ionic repulsion is important in limiting aggregation. Further, the binding of positively-charged materials to casein micelles, which accelerated aggregation, had little effect on the measured surface charge (Green, 1982). Despite this, it was shown that such binding can involve charge neutralization and it was suggested that it may result in the shielding of charged groups in the micelles, so increasing their hydrophobicity and favouring hydrophobic interactions between them. Since hydrophobic binding is endothermic, this idea accords with the marked increase in aggregation rate with rise in temperature, giving an activation energy (E_{act}) at moderate temperatures of close to $240 \, kJ \, mol^{-1}$ (calculated from Berridge, 1942; Mehaia & Cheryan, 1982). Such a high value for E_{act} is characteristic of a cooperative transition, in which many interacting components in a structure rearrange simultaneously (Tanford, 1973). Thus, micellar aggregation may involve several simultaneous chemical interactions, a view consistent with the appearance of intermicellar linkages on electron micrographs. The linkages appeared to be either wide bridges about 40 nm across, possibly consisting of a number of fine strands, or direct contact over a similar area (Green *et al.*, 1978) (Fig. 2). One can imagine that the formation of such complex links may require that the micelles approach one another in one of a few specific relative orientations in which the two areas of collision have complementary topographies. This provides a mechanism for steric limitations in micellar

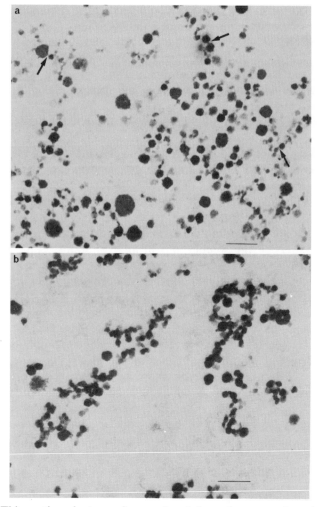

Fig. 2. Thin section electron micrographs of the early stages of casein micelle aggregation in skim milk: (a) at the clotting time; (b) at 3 times the clotting time. →, intramicellar bridges; scale bar = 400 nm.

aggregation. Probably, the individual interactions occurring do not only involve hydrophobic bonding. Dalgleish (1983) explained his observation that micellar aggregation was inhibited by increased ionic strength but accelerated by Ca^{2+} by proposing that specific ionic interactions occur between positively- and negatively-charged sites, such as are involved in the assembly of a bacteriophage.

At temperatures in excess of 50°C, the value of k_s approaches the diffusion-controlled limit, indicating the absence of an energy barrier to aggregation (Dalgleish, 1983). This may reflect an increased micellar hydrophobicity at higher temperatures, perhaps enabling the steric limitations to aggregation to be progressively overcome.

4.3. Curd Assembly

The later stages of aggregation, leading to assembly of the curd, have been studied microscopically. Initially, the casein micelles linked to form chains, which then gradually joined together to form a loose network (Green *et al.*, 1978) (Fig. 2). This became more extensive with time as the average size of the aggregates increased. The linkages between the micelles in the aggregates also appeared to become stronger with time. The proportion of bridges declined and it appeared that they might have contracted so bringing the micelles into contact. Eventually, the micelles began to fuse together. Probably, it is increases in the extent of aggregation and the strength of intermicellar linkages that lead to an increase in gel firmness.

The firmness of a rennet gel can continue to rise for at least 16 h under constant conditions (R. J. Marshall, unpublished work). This must reflect continuing aggregation of the casein micelles, which has been observed by microscopy over intervals of up to 24 h (Knoop, 1977). The network shrank with time, as thicker, shorter casein strands were formed. The casein particles came into closer contact, eventually leading to extensive fusion. Therefore, there is no need to suppose that micellar aggregation stops when the gel is cut. It seems much more probable that the process continues, causing contraction of the protein network, so that the whey is squeezed out but the fat is entrapped.

This view is confirmed by electron microscopic observations of changes in the structure of the curd during normal Chedder cheese-making (Kimber *et al.*, 1974). The casein micelles in the aggregates were extensively fused before cutting, but not enough to obscure their individualities (Fig. 3(a)). However, this process continued during scalding, forming continuous casein strands (Fig. 3(b)). In the later stages, the casein aggregated into progressively larger masses and there was pronounced curd shrinkage, causing distortion and possibly coalescence of the fat globules (Figs. 3(c) and (d)).

Steady aggregation of the casein was also observed to occur during cottage cheesemaking (Glaser *et al.*, 1980). In this instance, acidification was responsible for gel formation. There was a steady increase in the size of the casein particles during the manufacturing process, as the

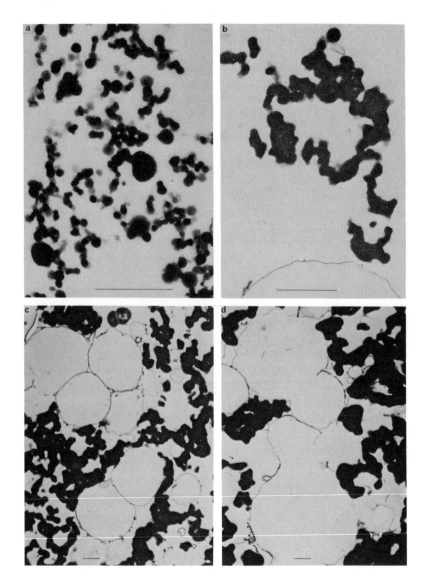

Fig. 3. Thin section electron micrographs of curd during the early stages of normal Cheddar cheesemaking. The stained particles and masses are casein and the stained lines are fat globule membranes. In the unprocessed specimen, the white areas would contain fat if enclosed by a membrane, or whey if not so enclosed. (a) Immediately before cutting, 40 min after renneting; (b) at maximum scald, 1·7 h after renneting; (c) 2·7 h after renneting; (d) at pitching, 3·5 h after renneting. Scale bar = 1 ɥm. (Modified from Kimber *et al.*, 1974.)

micelles aggregated into chains, then multiple strands and eventually coalesced into masses.

4.4. Mechanism of Syneresis and Subsequent Stages

Sorption, calorimetric (Rüegg *et al.*, 1974) and NMR measurements (Lelievre & Creamer, 1978) have shown that syneresis is not accompanied by a change in the degree of hydration of the casein. Thus, loss of the liquid entrapped by the casein network (amounting to some 90% of the original milk volume) must occur. As suggested above, this process is probably a continuation of casein aggregation. It is easy to envisage that the fusion of the particles in the casein network, the increase in the number of junction points and the shortening of the protein strands cause the entrapped liquid to be squeezed out steadily.

If this mechanism occurs, it follows that variables influencing the rate of formation of curd will also affect its syneresis to a similar extent and in the same direction. With one exception, it was found that all variables accelerating curd formation also accelerated syneresis and *vice versa* (Table 2). The exception is an increase in the protein content of the milk which increased the rate of development of curd rigidity but decreased the rate of whey loss. In fact, the higher rigidity of the

TABLE 2
Factors Influencing the Formation and Syneresis of Curd

Treatment of milk	Effect on curd formation	Effect on syneresis
Decrease pH 6·6 to 6·0	I[a]	I[b]
Raise temperature, 25 to 35°C	I[a]	I[b]
Increase rennet level	I[a]	I[b]
Add CaCl$_2$	I[a]	I[b]
Increase fat content	D[b]	D[b]
Increase protein content	I[c]	D[b]
Homogenization of fat	D[d]	D[d]
Cold storage	D[e]	D[e]
Psychrotroph growth	D[f]	D[f]
Increase pasteurization temperature	D[g]	D[h]

I, increased rate; D, decreased rate.
[a] Marshall *et al.* (1982); [b] Marshall (1982); [c] Storry & Ford (1982); [d] Davis (1965); [e] Knoop & Peters (1978); [f] Lelievre *et al.* (1978); [g] Marshall *et al.* (1978); [h] Schulz & Kley (1956).

curd appeared to be due to its higher protein or lower water content rather than a change in the rate of casein aggregation, since the latter was unaffected by the protein concentration (Green *et al.*, 1983). Similarly, the lower water content of this curd may have been responsible for its slower syneresis. Carrying the comparison further, those variables having the most marked affect on the formation of the curd, the pH value, temperature and pasteurization temperature, also had a large effect on the rate of syneresis. Conversely, those variables having a smaller effect on the rate of curd formation, rennet level, $CaCl_2$ addition, homogenization and cold storage of milk or psychrotroph growth, also affected syneresis to a lesser extent.

The view that whey loss is a continuation of curd formation is in accord with the very low value, about 0·1 mm of water, observed for the syneresis pressure (Van Dijk *et al.*, 1979), showing that syneresis is not due to the sudden application of large forces. It is also consistent with the relatively large influence of fat content on syneresis (Storry *et al.*, 1983), since fat globules enmeshed in the casein network could well act as 'plugs' inhibiting the flow of whey from the interstices.

Such evidence as is available suggests that curd fusion is influenced by the same factors as is milk coagulation. Difficulties in curd fusion were noted during Cheddar cheesemaking with homogenized milk, when curd formation and syneresis were both slowed (Green *et al.*, 1983). It has also been observed that high-heat-treated milks, which coagulate slowly, give curds which fuse poorly (R. J. Marshall, personal communication). However, other factors, such as the prevailing curd composition, temperature and pressure, may well influence curd fusion, and more work is needed before the relation between the initial and later stages of cheesemaking can be elucidated.

5. DEVELOPMENT OF CHEESE STRUCTURE AND TEXTURE

The evidence presented so far indicates that, once the aggregation of casein micelles is initiated, it continues as a major force during the whole cheesemaking process. It seems probable that the rate of aggregation of the casein is an important factor influencing both the composition and texture of the cheese. Before considering this in detail, the problem of describing cheese texture needs discussion.

5.1. Textural Evaluation of Cheese

A useful definition of the texture of a food must ultimately be expressed in sensory terms. However, instrumental measurements have certain advantages over sensory ones, in particular being easier to define and requiring fewer trained people. A start has only just been made in developing useful instrumental methods, but it is desirable, eventually, to be able to describe the textural characteristics of cheese in precise rheological terms, which, themselves, can be related to the composition and structure.

In sensory evaluation using a panel of trained people, the major difficulty is to select a number of easily-defined terms which describe all the textural characteristics of cheese. For hard and semi-hard cheeses, a combination of 5 terms has been found to be sufficient to describe texture. These are variously given as firmness or hardness, crumbliness or brittleness, granularity or lumpiness, springiness or rubberiness, and stickiness, adhesiveness or dryness (Brennan *et al.*, 1970; Lee *et al.*, 1978; Green *et al.*, 1981b).

Most instrumental methods currently used for testing cheese involve deforming the sample at a constant rate and measuring the force on it. The Instron Universal Testing Instrument seems to have been used most commonly. In some work, the load is applied in two cycles, along the lines given by Bourne (1978) for the texture testing of food. In general, the results for hard and semi-hard cheeses have been disappointing in that there tended to be less discrimination than with sensory measurements and really good correlations were rarely obtained. This might be due to inhomogeneities in the cheese, causing low reproducibility in instrumental measurements (Walstra & van Vliet, 1982). The most generally useful measurements appear to have been the force required to fracture the cheese and that required to give a defined deformation. Both, but especially the former, have tended to correlate with the firmness (Lee *et al.*, 1978; Green *et al.*, 1981b).

One problem with instrumental measurements is that a defined cheese sample tends to expand sideways so as to increase the area of contact with the probe and thus effectively decrease the load. To avoid this, Creamer & Olson (1982) used a probe carrying blocks of the same dimensions as the surface of the cheese sample so that the area of compression was kept constant. This enabled more clearly-defined force/compression curves to be obtained.

The description of cheese consistency in terms of viscoelastic parameters has been discussed by Walstra & van Vliet (1982), but little definitive experimental work has yet been done. However, a start has been made by Taneya *et al.* (1979) who developed equipment for applying an oscillating force of known frequency to samples of hard and semi-hard cheeses, enabling parameters describing the dynamic viscoelasticity to be determined.

5.2. Importance of Initial Structure of Coagulum

As well as the indirect influence on cheese texture by means of its effect on curd properties, the curd structure also has a more direct effect. The initial structure of the coagulum appears to be a major influence on the cheese structure. Curd made with bovine or pig pepsin had a more open protein network than that formed with rennet (Eino *et al.*, 1976), and similar differences were also observed between the corresponding Cheddar cheeses (Eino *et al.*, 1979). The protein network in curds prepared from milks concentrated by ultrafiltration (UF) became progressively coarser as the concentration factor of the milk increased (Green *et al.*, 1981b). As the curds were converted into Cheddar cheese, the structure altered but the structural differences between them were maintained (Fig. 4). The cheese structure was also found to influence the perceived texture directly. The firmness, crumbliness and granularity, measured by a sensory panel, and the firmness and force required to cause fracture, measured instrumentally, were all found to increase in proportion to the coarseness of the protein network. The major structural components influencing the texture were probably the size of the protein strands and the extent of the segregation of the fat and protein, both of which were larger in the cheeses from the more concentrated milks.

This experiment shows that the coagulum structure is influenced by the milk composition. The mineral content of the milk was also influential. Removal of more Ca, Mg, and phosphate than normal from the milk during ultrafiltration by prior acidification caused the protein network in the coagulum to be less coarse than that formed from unacidified milk of the same concentration factor (M. L. Green, unpublished work). The coagulum structure was also affected by homogenization of the fat in the milk, the protein network being finer than normal. When concentrated milk was also homogenized, the protein network in the coagulum was finer than that for non-homogenized milk concentrated to the same extent (Fig. 5).

An increased temperature for curd formation also caused the protein network in the coagulum to be coarser, and a reduced temperature

Fig. 4. Scanning electron micrographs of curds at different stages during Cheddar cheesemaking with milks concentrated by ultrafiltration. (a) and (d) at cutting; (b) and (e) at maximum scald; (c) and (f) at pressing. (a), (b) and (c) from unconcentrated milk; (d), (e) and (f) from milk concentrated 4-fold. Scale bar = 5 μm. (Modified from Green *et al.*, 1981b.)

caused the network to be less coarse (M. L. Green, unpublished work). Since a rise in temperature accelerates casein aggregation, it is suggested that faster aggregation may be the cause of the coarser protein network. Van Dijk (1982) reached a similar conclusion. He found that curd which had formed more rapidly, for instance by raising

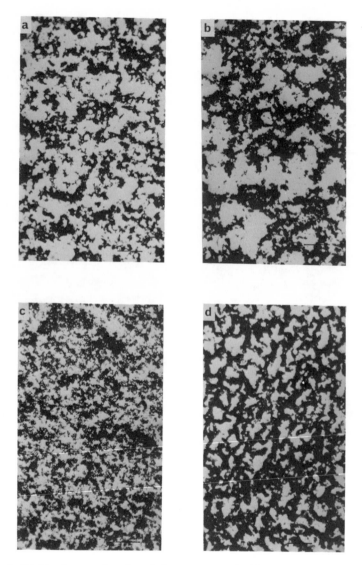

Fig. 5. Light micrographs of curds from processed milks. Samples of curd were taken 40 min after rennet addition. Milks: (a) control; (b) concentrated 2·5-fold by ultrafiltration; (c) homogenized; (d) concentrated 2·5-fold by ultrafiltration in a plant causing homogenization. Scale bar = 20 μm.

the temperature, tended to be more permeable, indicating that the protein network was more open. The same mechanism also explains the effects of mineral removal and homogenization, both of which decrease the casein aggregation rate. The effect of homogenization probably results from the linkage of many casein micelles to fat particles, increasing the size of the primary aggregating particles. Thus, the average diffusion coefficient of the particles and, consequently, the rate of aggregation, will be reduced (eqn (1)).

Knoop & Peters (1975) suggested a mechanism by which such an influence could occur. The energy barrier to linkage (W) of two chains of particles is smallest if the chains add end-to-end, and greatest if they add laterally (Thomas & McCorkle, 1971). Similar considerations apply to the addition of single particles to existing chains. Therefore, when W is higher, which is generally when aggregation is slower, there will be a greater tendency for end-to-end addition, so the chains of particles will tend to be thinner and the network will be of finer mesh.

This mechanism could explain the effects of mineral content of the milk, temperature and homogenization on coagulum structure. The effects of concentration may result from the proportionate reduction in the mean free path of micelles from about 4 particle diameters in normal milk. Therefore, interacting particles will be more influenced by other micelles in the concentrated milks. This will tend to reduce the directional forces exerted by any one particle or chain and so increase the probability of addition occurring laterally.

The structure of the coagulum also affects its properties. The fat losses were higher during cheesemaking with bovine pepsin than with rennet, probably due to the more open structure of the pepsin curds (Stanley & Emmons, 1977). Curds formed from UF-concentrated milks became progressively less effective at retaining both fat and moisture as the protein network became coarser (Green *et al.*, 1981a). Curds from homogenized milk, having a finer protein network than normal, retained excessive moisture and did not fuse well (Green *et al.*, 1983). As expected from this, homogenization of concentrated milk gave curd retaining more moisture and fat and yielding cheese with somewhat improved textural properties.

In a number of less detailed studies, a relation between milk treatment or composition and cheese structure or texture has also been noted. Homogenization of cream in milk affected both the structure and texture of Camembert cheese (Prokopek *et al.*, 1976) and gave a softer, smoother, more elastic Cheddar cheese (Emmons *et al.*, 1980).

The texture of Camembert cheese made from UF-concentrated milk could be improved by heat treatment and lowering the temperature of curd formation (Prokopek *et al.*, 1976). The poorer fusion of curds from high-heat-treated milks tended to cause a crumbly, open texture (R. J. Marshall, personal communication). Compared with full-fat milks, those with reduced fat gave Cheddar cheeses which were more crumbly, but also firmer and more elastic (Emmons *et al.*, 1980), and Domiati cheeses which were more rubbery and less smooth (Kerr *et al.*, 1981).

5.3. Later Cheesemaking Stages

Workers in New Zealand have taken the view that there is a direct link between the structure of a cheese and its composition (Lawrence *et al.*, 1983). They opine that the cheese structure depends primarily on the acidity of the curd at whey draining. This determines the amounts of Ca and phosphate remaining in the curd and hence the extent of aggregation of the casein in the final cheese. While the composition of the cheese undoubtedly influences both the aggregation of the protein at the submicellar level (Hall & Creamer, 1972) and the cheese texture (Chen *et al.*, 1979), it is difficult to see how the proposed mechanism could have much influence on the distribution of fat within the protein network.

However, there is evidence for some influence of the later cheese-making stages on the curd structure and, thus, the final cheese texture. The flow during cheddaring, possibly aided by the scald, caused some aggregation of the fat globules (Taranto *et al.*, 1979) and orientation of the protein in Cheddar cheese (Kalab, 1977). Vacuum pressing markedly reduced openness in Cheddar cheese (Hoglund *et al.*, 1972) and also rendered the milled curd junctions almost invisible. Otherwise, it was at these junctions that openness tended to occur (Lowrie *et al.*, 1982).

Higher heating rates during cottage cheese manufacture increased the firmness of the curd particles at a constant moisture level (Chua & Dunkley, 1979). Increase in cooking time rendered processed Cheddar cheese firmer and more elastic and led to smaller fat particles (Rayan *et al.*, 1980).

5.4. Factors Affecting Cheese Ripening

During the ripening of many types of cheeses, especially during the earlier stages, the protein matrix is converted to a smoother, more homogeneous structure and softening occurs (Stanley & Emmons, 1977; De Jong, 1978). This is probably due to proteolysis, the extent of which correlated with softening in a soft cheese (De Jong, 1977). In that instance, the hydrolysis of α_{s1}-casein by residual coagulant was probably most important. This process results in loss of capacity for hydrophobic interactions, which could well cause the protein network to become weaker (Creamer *et al.*, 1982). However, in appropriate types 'of cheeses, mould development is a major factor determining proteolysis, since this tends to occur fastest nearest the fungal hyphae (Knoop & Peters, 1971).

A number of factors may influence the rate of proteolysis in cheeses. It was dependent on the moisture content in one type of soft cheese (De Jong, 1978). The salt-in-moisture level in Cheddar cheese also had a marked effect on the rate of proteolysis (Pearce, 1982), and the salt concentration was correlated with firmness in Dutch cheese (Ramanauskas, 1978). Increase in the ripening temperature also accelerated proteolysis in cheese and could affect the rheological properties (Davis, 1965).

The extent of protein breakdown is probably much more important than the type in giving the desired texture in ripe cheese. Most likely, it is by influences on the rate of proteolysis that the composition of cheese affects the textural properties. The important factors, moisture-in-non-fat-solids, salt-in-moisture and pH (Lawrence & Gilles, 1980) control the concentrations of water and salt and the acidity, respectively, in the cheese. These all have marked effects on enzymic activity.

The composition of the cheese may also have a direct influence on the texture. The moisture content had a greater effect than the water-soluble N content, a measure of the extent of proteolysis, on the hardness of a range of cheeses (Rüegg *et al.*, 1980). The fat content was also influential in the same regard. These factors probably act by influencing the flexibility of the protein network and the ability of areas of fat to move in relation to the matrix, respectively.

6. FACTORS AFFECTING THE CHEESEMAKING POTENTIAL OF MILKS

Milk composition can influence coagulum structure, which is then an important determinant of both cheese texture and composition (Fig. 6). The basis for these links may be that the cheesemaking stages

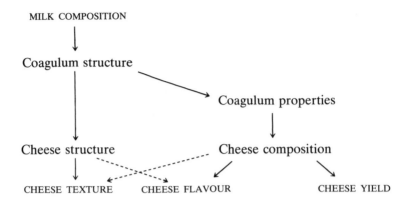

Fig. 6. Means by which milk composition affects yields and properties of cheese.

between rennet treatment and pressing appear to be driven by the continuing tendency of casein to aggregate. This could also explain the observation that the same factors cause slow coagulation, weak coagula and excessive moisture retention (Davis, 1965). Moreover, coagulation time, coagulum strength and moisture retention are affected by all the chemical factors influencing the non-flavour aspects of cheese quality, i.e. composition and texture. For instance, weaker coagula tend to yield Cheddar cheeses containing higher moisture levels (Chapman & Burnett, 1972).

6.1. Treatment
The effects of homogenization and concentration of milk on coagulum structure and properties and on cheese composition and texture have already been discussed.

High heat treatment of milk causes interaction of the whey proteins, especially β-lactoglobulin, with the outside of the casein micelles (Davies *et al.*, 1978). As a result the milk coagulates more slowly than normal and forms weaker, more fragile coagula. The curd retains excessive moisture and fusion of the particles is less complete than normal (R. J. Marshall, personal communication). In cottage cheese-making with such milks, it was noted that fines formation and curd breakage tended to occur (Emmons *et al.*, 1982) and the product had a soft, weak body and was high in moisture (Kosikowski, 1982).

Storage of milk at low temperatures can cause slower coagulation, formation of a weak coagulum and prolonged cheesemaking time (Reimerdes *et al.*, 1977; Ali *et al.*, 1980). This appears to be due to solubilization of Ca, phosphate and casein from the casein micelles, since this was reversed and the cheesemaking properties of the milk were restored by heating at 60°C. It appears that the effects of cooling can be avoided by thermization, heating at 60–70°C for 10–15 s, before storage (Johnston *et al.*, 1981).

6.2. Composition

Milk composition is influenced by genetic, physiological and husbandry factors. In practice, the suitability of milks for cheesemaking are probably affected most by the health, feed and calving pattern of a herd, and the season. The presence of mastitis is a major problem, causing the formation of weak coagula and associated defects. Mastitic milks differ from normal milks in having a high pH and high concentrations of Na^+, chloride and soluble protein and low Ca, phosphate and casein contents (Kitchen, 1981), the last perhaps depleted further by high levels of proteolytic enzymes (Anderson & Andrews, 1977). Probably the levels of both salts and casein are influential, since the coagulum strength of mastitic milk can be improved, but not brought up to the normal level, by dialysis against normal milk (Erwin *et al.*, 1972).

So as to maximize the variations between milks, experimental studies of the relation between milk composition and coagulating properties have generally used milks of individuals or small groups of cows. However, normally the concentrations of a number of components vary together (Table 3) and this has rendered the interpretation of the results difficult. The major factor affecting the coagulation of a milk by rennet appeared to be its pH (White & Davies, 1958; Morrissey *et al.*, 1981). The total casein concentration may also have been important (Storry *et al.*, 1983).

More definitive results were obtained when the compositions of milk were manipulated *in vitro* (Flüeler & Puhan, 1979). The differences in coagulation time disappeared when milks were diafiltered so as to remove all salts except Ca, which was held at a fixed concentration. This clearly shows that the level and distribution of Ca are important in determining coagulability.

Many of the compositional characteristics of poorly coagulating milks, such as high levels of Na^+, soluble citrate and phosphate and low levels of soluble Ca (Table 3), tend to either indicate or cause a

TABLE 3
Compositional Characteristics of Poorly-coagulating Milks

Characteristics	References
Low casein and Ca, high pH, Na^+ and chloride	Davis (1965)
High proportion of α_{s1}-casein, low casein-bound Ca and soluble Ca/(phosphate + citrate)	Flüeler (1978)
High α_{s1}-casein/β-casein, low (Ca + Mg)/(phosphate + citrate)	Storry *et al.* (1983)
High soluble citrate and phosphate, low soluble Ca and Mg	Flüeler & Puhan (1979)
Low total citrate and soluble Ca	Morrissey *et al.* (1981)
Low colloidal calcium phosphate–citrate	Pyne (1962)

low degree of Ca binding by casein (Pyne, 1962). Further, α_{s1}-casein has more binding sites for Ca than does β-casein (Parker & Dalgleish, 1981). Therefore, milks with a higher ratio of α_{s1}-casein/β-casein require a higher level of bound Ca to reach the normal extent of saturation of the binding sites by Ca. Thus, it appears that milks may coagulate poorly when the degree of saturation of the binding sites on casein by Ca is low.

High levels of association of Ca with casein, low pH and high casein concentration all tend to increase both the tendency of casein to form a coagulum and the proportion of casein in micellar form (Lin *et al.*, 1972). This suggests that these two properties may be linked, and that the factors favouring the association of casein to form micelles also favour the association of para-casein to form a coagulum. A number of polycations which bind to casein have already been shown to facilitate both these reactions (Green, 1982).

The mechanism suggested for the influence of the composition of milk on its coagulation can only rest on qualitative arguments at present. However, two computer models have been proposed which enable the distribution of ions among the various possible complexes and levels of ionization to be determined in ultrafiltrates at any pH value and temperature (Holt *et al.*, 1981; Lyster, 1981). Application of these may assist identification of the actual components present in the milk which affect its cheesemaking properties. However, it is not yet possible to describe the milk system completely because the binding of ions to the proteins is not known accurately. This requires precise knowledge of the composition of micellar calcium phosphate, the first determinations of which are now available (Holt, 1982). Thus, it should soon be possible to extend the computer programs to include the interactions of milk salts with proteins.

7. CONCLUSION

Our knowledge of the early stages of cheesemaking, especially milk coagulation, has increased considerably in recent years, although the precise details of the mechanism of micellar interaction are still unclear. Our understanding of the later stages, and the influence of variables upon these, is much more sketchy. A major factor limiting progress has probably been the lack of theoretically-sound, reproducible methods for measuring aspects of curd formation and behaviour. This problem is being addressed with respect to curd formation and syneresis, but little attention has yet been paid to curd fusion and salt uptake.

Interest is now being directed towards the use of a wider range of milks, especially homogenized, concentrated or heated ones, for cheesemaking. This arises because the milks are products of other desirable processes or because of the possibility of raising efficiency by increasing yields or throughput. Further, the composition of raw milk may change over small time intervals as farmers seek to maximize profitability by manipulating feed components. These may all influence the cheesemaking properties, potentially affecting cheese quality through compositional and textural changes, and yields, through reduced retention of components in the cheese. Generally, cheese composition, texture and yield are all influenced simultaneously, so all need to be considered when changes in the milk composition or processing are contemplated.

REFERENCES

ALI, A. E., ANDREWS, A. T. & CHEESEMAN, G. C. (1980) *J. Dairy Res.* **47,** 371.
ANDERSON, M. & ANDREWS, A. T. (1977) *J. Dairy Res.* **44,** 223.
ANDRÉN, A., BJÖRCK, L. & CLAESSON, O. (1981) *Netherlands Milk Dairy J.* **35,** 365.

ANIFANTAKIS, E. & GREEN, M. L. (1980) *J. Dairy Res.* **47**, 221.
BEEBY, R. (1979) *New Zealand J. Dairy Sci. Technol.* **14**, 1.
BERRIDGE, N. J. (1942) *Nature* **149**, 194.
BOURNE, M. C. (1978) *Food Technol.* **32**(7), 62.
BRENNAN, J. G., JOWITT, R. & MUGHSI, O. A. (1970) *J. Texture Studies* **1**, 167.
BURGESS, K. J. (1978) *Irish J. Food Sci. Technol.* **2**, 129.
CHAPLIN, B. & GREEN, M. L. (1980) *J. Dairy Res.* **47**, 351.
CHAPLIN, B. & GREEN, M. L. (1982) *J. Dairy Res.* **49**, 631.
CHAPMAN, H. R. & BURNETT, J. (1972) *Dairy Ind.* **37**, 207.
CHEN, A. H., LARKIN, J. W., CLARK, C. J. & IRWIN, W. E. (1979) *J. Dairy Sci.* **62**, 901.
CHUA, T. E. H. & DUNKLEY, W. L. (1979) *J. Dairy Sci.* **62**, 1216.
COLLIN, J.-C., MUSET DE RETTA, G. & MARTIN, P. (1982) *J. Dairy Res.* **49**, 221.
CORRADINI, C. & DIECI, E. (1978) *Scienza e Tecnica Lattiero-Casearia* **29**, 382.
CREAMER, L. K. & OLSON, N. F. (1982) *J. Food Sci.* **47**, 631.
CREAMER, L. K., ZOERB, H. F., OLSON, N. F. & RICHARDSON, T. (1982) *J. Dairy Sci.* **65**, 902.
DALGLEISH, D. G. (1982) In *Developments in Dairy Chemistry—I: Proteins*, P. F. Fox, Ed. Applied Science Publishers, London and New York, p. 157.
DALGLEISH, D. G. (1983) *J. Dairy Res.* **50**, 331.
DALGLEISH, D. G., BRINKHUIS, J. & PAYENS, T. A. J. (1981) *European J. Biochem.* **119**, 257.
DARLING, D. F. & VAN HOOYDONK, A. C. M. (1981) *J. Dairy Res.* **48**, 189.
DAVIES, F. L., SHANKAR, P. A., BROOKER, B. E. & HOBBS, D. G. (1978) *J. Dairy Res.* **45**, 53.
DAVIS, J. G. (1965) *Cheese*, vol. 1. J. & A. Churchill Ltd, London.
DE JONG, L. (1977) *Netherlands Milk Dairy J.* **31**, 314.
DE JONG, L. (1978) *Netherlands Milk Dairy J.* **32**, 1, 15.
DE KONING, P. J. (1980) *IDF Bulletin*, Doc. 126, 11.
DE KONING, P. J., VAN ROOIJEN, P. J. & VISSER, S. (1978) *Netherlands Milk Dairy J.* **32**, 232.
DOUILLARD, R. (1973) *J. Texture Studies* **4**, 158.
EINO, M. F., BIGGS, D. A., IRVINE, D. M. & STANLEY, D. W. (1976) *J. Dairy Res.* **43**, 113.
EINO, M. F., BIGGS, D. A., IRVINE, D. M. & STANLEY, D. W. (1979) *Canadian Inst. Food Sci. Technol. J.* **12**, 149.
EMMONS, D. B., REISER, B., GIROUX, R. N. & STANLEY, D. W. (1976) *Canadian Inst. Food Sci. Technol. J.* **9**, 189.
EMMONS, D. B., KALAB, M., LARMOND, E. & LOWRIE, R. J. (1980) *J. Texture Studies* **11**, 15.
EMMONS, D. B., BECKETT, D. C. & MODLER, H. W. (1982) *Proc. 21st Int. Dairy Congress, Moscow* **1**(1), 407.
ERNSTROM, C. A. (1974) In *Fundamentals of Dairy Chemistry*, B. H. Webb, A. H. Johnson and J. A. Alford, Eds, 2nd edn. Avi Publ. Co., Conn., p. 662.

ERWIN, R. E., HAMPTON, O. & RANDOLPH, H. E. (1972) *J. Dairy Sci.* **55**, 298.

FLÜELER, O. (1978) *Schweizerische Milchwirtschaftliche Forschung* **7**, 45.

FLÜELER, O. & PUHAN, Z. (1979) *Schweizerische Milchwirtschaftliche Forschung* **8**, 49.

FOLTMANN, B. (1981) *Essays in Biochem.* **17**, 52.

GARNOT, P., TOULLEC, R., THAPON, J.-L., MARTIN, P., HOANG, M.-T., MATHIEU, C. M. & RIBADEAU DUMAS, B. (1977) *J. Dairy Res.* **44**, 9.

GLASER, J., CARROAD, P. A. & DUNKLEY, W. L. (1980) *J. Dairy Sci.* **63**, 37.

GOODWIN, J. W. & KHIDHER, A. M. (1976) *Colloid Interface Sci.* **4**, 529.

GORDIN, S. & ROSENTHAL, I. (1978) *J. Food Protection* **41**, 684.

GREEN, M. L. (1977) *J. Dairy Res.* **44**, 159.

GREEN, M. L. (1980) *Milk Ind.* **82**(3), 25.

GREEN, M. L. (1982) *J. Dairy Res.* **49**, 87, 99.

GREEN, M. L. & MORANT, S. V. (1981) *J. Dairy Res.* **48**, 57.

GREEN, M. L., HOBBS, D. G., MORANT, S. V. & HILL, V. A. (1978) *J. Dairy Res.* **45**, 413.

GREEN, M. L., GLOVER, F. A., SCURLOCK, E. M. W., MARSHALL, R. J. & HATFIELD, D. S. (1981a) *J. Dairy Res.* **48**, 333.

GREEN, M. L., TURVEY, A. & HOBBS, D. G. (1981b) *J. Dairy Res.* **48**, 343.

GREEN, M. L., MARSHALL, R. J. & GLOVER, F. A. (1983) *J. Dairy Res.* **50**, 341.

HALL, D. M. & CREAMER, L. K. (1972) *New Zealand J. Dairy Sci. Technol.* **7**, 95.

HARBOE, M. K. (1981) *Netherlands Milk Dairy J.* **35**, 367.

HARRIS, T. J. R., LOWE, P. A., LYONS, A., THOMAS, P. G., EATON, M. A. W., MILLICAN, T. A., PATEL, T. P., BOSE, C. C., CAREY, N. H. & DOEL, M. T. (1982) *Nucleic Acids Res.* **10**, 2177.

HATFIELD, D. S. (1981) *J. Soc. Dairy Technol.* **34**, 139.

HOGLUND, G. F., FRYER, T. F. & GILLES, J. (1972) *New Zealand J. Dairy Sci. Technol.* **7**, 150.

HOLT, C. (1982) *J. Dairy Res.* **49**, 29.

HOLT, C., DALGLEISH, D. G. & JENNESS, R. (1981) *Anal. Biochem.* **113**, 154.

HUŠEK, V. & DĚDEK, M. (1981) *Netherlands Milk Dairy J.* **35**, 302.

JOHNSTON, D. E., MURPHY, R. J., GILMOUR, A. & MACELHINNEY, R. S. (1981) *Milchwissenschaft* **36**, 748.

KALAB, M. (1977) *Milchwissenschaft* **32**, 449.

KALAB, M., LOWRIE, R. J. & NICHOLS, D. (1982) *J. Dairy Sci.* **65**, 1117.

KERR, T. J., WASHAM, C. J., EVANS, A. L. & TODD, R. L. (1981) *J. Food Protection* **44**, 496.

KIMBER, A. M., BROOKER, B. E., HOBBS, D. G. & PRENTICE, J. H. (1974) *J. Dairy Res.* **41**, 389.

KITCHEN, B. J. (1981) *J. Dairy Res.* **48**, 167.

KNOOP, A.-M. (1977) *Deutsche Milchwirtschaft* **28**, 1154.

KNOOP, A.-M. & PETERS, K.-H. (1971) *Milchwissenschaft* **26**, 193.

KNOOP, A.-M. & PETERS, K.-H. (1975) *Kieler Milchwirtschaftliche Forschungberichte* **27**, 227.

KNOOP, A.-M. & PETERS, K.-H. (1978) *Proc. 20th Int. Dairy Congress, Paris,* 808.

KOBAYASHI, H., KUSAKABE, I. & MURAKAMI, K. (1982) *Anal. Biochem.* **122,** 308.

KOSIKOWSKI, F. V. (1982) *J. Dairy Sci.* **65** (supplement), 73.

KOWALCHYK, A. W. & OLSON, N. F. (1978) *J. Dairy Sci.* **61,** 1375.

LAWRENCE, R. C. & GILLES, J. (1980) *New Zealand J. Dairy Sci. Technol.* **15,** 1.

LAWRENCE, R. C., GILLES, J. & CREAMER, L. K. (1983) *New Zealand J. Dairy Sci. Technol.* **18,** 175.

LEE, L.-H., IMOTO, E. M. & RHA, C. (1978) *J. Food Sci.* **43,** 1600.

LELIEVRE, J. & CREAMER, L. K. (1978) *Milchwissenschaft* **33,** 73.

LELIEVRE, J., KELSO, E. A. & STEWART, D. B. (1978) *Proc. 20th Int. Dairy Congress, Paris,* 760.

LIN, S. H. C., LEONG, S. L., DEWAN, R. K., BLOOMFIELD, V. A. & MORR, C. V. (1972) *Biochemistry* **11,** 1818.

LOWRIE, R. J., KALAB, M. & NICHOLS, D. (1982) *J. Dairy Sci.* **65,** 1122.

LYSTER, R. L. J. (1981) *J. Dairy Res.* **48,** 85.

MARSHALL, R. J. (1982) *J. Dairy Res.* **49,** 329.

MARSHALL, R. J., CHAPMAN, H. R. & GREEN, M. L. (1978) *Proc. 20th Int. Dairy Congress, Paris,* 805.

MARSHALL, R. J., HATFIELD, D. S. & GREEN, M. L. (1982) *J. Dairy Res.* **49,** 127.

MARTIN, P., COLLIN, J.-C., GARNOT, P., RIBADEAU DUMAS, B. & MOC-QUOT, G. (1981) *J. Dairy Res.* **48,** 447.

MATHESON, A. R. (1981) *New Zealand J. Dairy Sci. Technol.* **16,** 33.

MCMAHON, D. J. & BROWN, R. J. (1982) *J. Dairy Sci.* **65,** 1639.

MEHAIA, M. A. & CHERYAN, M. (1982) *J. Dairy Sci.* **65** (supplement), 56.

MORRISSEY, P. A., MURPHY, M. F., HEARN, C. M. & FOX, P. F. (1981) *Irish J. Food Sci. Technol.* **5,** 117.

PARKER, T. G. & DALGLEISH, D. G. (1981) *J. Dairy Res.* **48,** 71.

PAYENS, T. A. J. (1977) *Biophys. Chem.* **6,** 263.

PEARCE, K. N. (1979) *New Zealand J. Dairy Sci. Technol.* **14,** 233.

PEARCE, K. N. (1982) *Proc. 21st Int. Dairy Congress, Moscow* **1**(1), 519.

PHELAN, J. A. & O'BRIEN, M. F. (1982) *Proc. 21st Int. Dairy Congress, Moscow* **1**(2), 514.

PROKOPEK, D., KNOOP, A. M. & BUCHHEIM, W. (1976) *Kieler Milchwirtschaftliche Forschungsberichte* **28,** 245.

PYNE, G. T. (1962) *J. Dairy Res.* **29,** 101.

RAMANAUSKAS, R. (1978) *Proc. 20th Int. Dairy Congress, Paris,* 265.

RAYAN, A. A., KALAB, M. & ERNSTROM, C. A. (1980) *Scanning Electron Microscopy* **3,** 635.

REIMERDES, E.-H., PEREZ, S. J. & RINGQVIST, B. M. (1977) *Milchwissenschaft* **32,** 154.

ROTHE, G. A. L., AXELSEN, N. H., JØHNK, P. & FOLTMANN, B. (1976) *J. Dairy Res.* **43,** 85.

RÜEGG, M., LUSCHER, M. & BLANC, B. (1974) *J. Dairy Sci.* **57,** 387.

RÜEGG, M. EBERHARD, P., MOOR, U., FLÜCKIGER, E. & BLANC, B. (1980). *Schweizerische Milchwirtschaftliche Forschung* **9,** 3.

SCHULZ, M. E. & KLEY, W. (1956) *Milchwissenschaft* **11**, 116.
SCOTT BLAIR, G. W. (1971) *Rheologica Acta* **10**, 316.
SCOTT BLAIR, G. W. & BURNETT, J. (1958) *J. Dairy Res.* **25**, 297.
SHANLEY, R. M. & JAMESON, G. W. (1981) *Australian J. Dairy Technol.* **36**, 107.
STADHOUDERS, J., HUP, G. & VAN DER WAALS, C. B. (1977) *Netherlands Milk Dairy J.* **31**, 3.
STANLEY, D. W. & EMMONS, D. B. (1977) *Canadian Inst. Food Sci. Technol. J.* **10**, 78.
STANLEY, D. W., EMMONS, D. B., MODLER, H. W. & IRVINE, D. M. (1980) *Canadian Inst. Food Sci. Technol. J.* **13**, 97.
STORRY, J. E. & FORD, G. D. (1982) *J. Dairy Res.* **49**, 343.
STORRY, J. E., GRANDISON, A. S., MILLARD, D., OWEN, A. J. & FORD, G. D. (1983) *J. Dairy Res.* **50**, 215.
TAKAHASHI, F., REIMERDES, E.-H. & KLOSTERMEYER, H. (1978) *Proc. 20th Int. Dairy Congress, Paris,* 447.
TANEYA, S., IZUTSU, T. & SONE, T. (1979) In *Food Texture and Rheology,* P. Sherman, Ed. Academic Press Inc, London.
TANFORD, C. (1973) *The Hydrophobic Effect: Formation of Micelles and Biological Membranes.* John Wiley, New York.
TANG, J., JAMES, M. N. G., HSU, I. N., JENKINS, J. A. & BLUNDELL, T. L. (1978) *Nature* **271**, 618.
TARANTO, M. V., WAN, P. J., CHEN, S. L. & RHEE, K. C. (1979) *Scanning Electron Microscopy* **3**, 273.
TAYLOR, M. J., OLSON, N. F. & RICHARDSON, T. (1979) *Process Biochem.* **14**(2), 10.
THOMAS, I. L. & MCCORKLE, K. H. (1971) *J. Colloid Interface Sci.* **36**, 110.
THOMASOW, J. (1980) *Milchwissenschaft* **35**, 212.
THUNELL, R. K., DUERSCH, J. W. & ERNSTROM, C. A. (1979) *J. Dairy Sci.* **62**, 373.
TOKITA, M., HIKICHI, K., NIKI, R. & ARIMA, S. (1982) *Biorheology* **19**, 209.
VANDERHEIDEN, G. (1976) *CSIRO Food Res. Quarterly* **36**, 45.
VAN DIJK, H. J. M. (1982) Thesis, University of Wageningen.
VAN DIJK, H. J. M., WALSTRA, P. & GUERTS, T. J. (1979) *Netherlands Milk Dairy J.* **33**, 60.
VAN HOOYDONK, A. C. M. & OLIEMAN, C. (1982) *Netherlands Milk Dairy J.* **36**, 153.
VISSER, S. (1981) *Netherlands Milk Dairy J.* **35**, 65.
WAHBA, A. & EL-ABBASSY, F. (1981) *Egyptian J. Dairy Sci.* **9**, 11.
WALSTRA, P. (1979) *J. Dairy Res.* **46**, 317.
WALSTRA, P. & VAN VLIET, T. (1982) *IDF Bulletin,* Doc. 153, 22.
WALSTRA, P., BLOOMFIELD, V. A., WEI, G. J. & JENNESS, R. (1981) *Biochimica et Biophysica Acta* **669**, 258.
WHITE, J. C. D. & DAVIES, D. T. (1958) *J. Dairy Res.* **25**, 267.
ZANNONI, M., MONGARDI, M. & ANNIBALDI, S. (1981) *Scienza e Tecnica Lattiero–Casearia* **32**, 153.
ZVIEDRANS, P. & GRAHAM, E. R. B. (1981) *Australian J. Dairy Technol.* **36**, 117.

Chapter 2

Taxonomy and Identification of Bacteria Important in Cheese and Fermented Dairy Products

ELLEN I. GARVIE

National Institute for Research in Dairying, Shinfield, Reading, UK

1. INTRODUCTION

Man has herded animals for many centuries and used cheese and fermented milks. The traditional cheese making processes must have evolved gradually and, at the same time, the bacteria most suited to grow under the conditions used would be selected. The cultures found in traditional fermentations were finally isolated, purified and removed to laboratories about 100 years ago when the science of microbiology was new.

The large scale use of starter cultures is a modern development; commercial conditions demand that strains can be identified and that they can be maintained to make uniform products. Even so, some important bacteria in cheese are not deliberately added but develop from chance contamination, and some of these are seldom found in other habitats. Cheese and fermented dairy products give a specialized environment highly selective for certain bacteria. Strains adapted to this environment are, in some species, different from strains of the same species living away from milk and milk products. The differences can be important in commercial practice. The 'dairy' strains need to be identified, but for the taxonomist, the differences between the wild and dairy strains is of less importance. Identification for the practical dairyman and classification for the purpose of taxonomy are not necessarily the same.

A logical systematic scheme based on evolution has been developed by botanists and zoologists aided by fossil records. The microbiologist has no fossils to aid taxonomy and classical classification has been based on morphology and the biochemical ability of the culture (usually growing) under standard conditions. In the last few decades this has changed. The chemical composition of fundamental cell components can now be studied and used to unravel the relationship between different bacteria. The microbiologist, like taxonomists in other disciplines, is beginning to base classification logically on evolution. In some cases the newer knowledge has confirmed the older classification based on superficial properties but in other instances changes have been necessary. Where these changes cut across long established concepts they are only slowly accepted.

This chapter reviews modern chemical, enzymological and physiological techniques and relates them to current thinking on the classification of dairy bacteria. Particular attention is given to the effects of the milk environment on the evolution of dairy lactic-acid bacteria.

2. APPROACHES TO BACTERIAL TAXONOMY AND THEIR INFLUENCE ON THE CLASSIFICATION OF DAIRY BACTERIA

2.1. Nucleic Acid Composition and Homology

2.1.1. Ribonucleic Acid

Ribonucleic acid (RNA) and deoxyribonucleic acid (DNA) are being used increasingly in taxonomy. Information about RNA divides bacteria into larger groups while DNA gives information at a species level. Woese & Fox (1977a) and Fox *et al.* (1980) discuss the concept of evolution based on RNA, and the latter paper gives a schematic representation of the major lines of prokaryote descent. The workers at Illinois developed their schemes from information on 16s ribosomal RNA sequencing which is summarized by Woese & Fox (1977b). Representatives of a large number of bacterial genera have been studied to give an overall picture. It is believed that ribosomal RNA is a slowly evolving molecule which allows the detection of the relationship between very distant species (Woese & Fox, 1977b) and the information obtained predates that from the fossil record (Woese & Fox, 1977a). Knowledge is still scanty and any scheme based on it could well be changed as more information becomes available.

The scheme of prokaryote descent mentioned above has been augmented by Kandler (1982) who gives an interesting phylogenetic 'bush', adding information on cell wall structure to that from 16s RNA studies.

Cell wall components will be discussed later as they can also be used to assist taxonomy at a different level from RNA. Commercial interests require simple quick and reliable methods for rapid identification of species and strains. Fundamental taxonomy, studying RNA and looking back at evolution millions of years ago, may seem irrelevant to the practical dairyman. However, the needs of industry will be best met when the methods of identification used are soundly based on an evolutionary scheme.

Ribosomal RNA is readily isolated from the bacterial cell and purified (Moore & McCarthy, 1967). It will hybridize with single stranded DNA and various techniques are available for determining the formation of the hybrids (Gillespie, 1968). The heat stability of the DNA/RNA hybrid expressed by its melting temperature ($T_{m(e)}$) has been used in several studies of Gram-negative bacteria and 'coryneforms' (Gillis & DeLey, 1980). Limited studies have been made of some streptococci (Garvie & Farrow, 1981; Kilpper-Balz *et al.*, 1982) but not of other lactic-acid bacteria. The diversity between species of streptococci appears to be greater than that between several genera of Gram-negative bacteria. Information is too scanty for firm conclusions to be drawn but this approach to taxonomy could give much valuable information to supplement the ribosomal RNA sequencing.

2.1.2. Deoxyribonucleic Acid

DNA was used before RNA in the classification of bacteria. The composition of DNA expressed as % guanine + cytosine (% G + C) is now known for most species. Within a species the % G + C falls within a narrow range for different strains and a natural genus will contain species with % G + C varying by only a few percent, but an exact acceptable range has not been established. Genera like the lactobacilli with a wide range of % G + C (36–52) require further study to see if they are a natural genus. The clear separation of micrococci (% G + C 66–75) from staphylococci (% G + C 30–40) is well established and this big difference is not difficult to understand as the micrococci and staphylococci are far separated in evolution (Kandler, 1982). Several techniques are available for estimating % G + C. Heat denaturation and estimation of the melting temperature (T_m) is widely used now that it can be related to actual chemical determination of guanine and cytosine content.

At first it seemed that plasmids posed a problem in work with DNA. Only double stranded high molecular weight DNA is used in estimation of % G + C and plasmids, unless they are incorporated in the genome, will be lost during purification procedures. In any case plasmid DNA will only be a small part of the total in a bacterial cell and will not affect the estimated % G + C. Plasmids move from strain to strain but the extent to which this takes place outside controlled laboratory conditions is probably small. The stability of RNA and of other fundamental cell constituents indicates that bacterial species exist and are stable. Reanney *et al.* (1983) discuss genetic interactions amongst bacteria and conclude that bacterial genomes remain stable in nature. Bacteria probably do not evolve faster than other types of organism.

Hydridization between single stranded DNA from different strains is an excellent measure of the relationship between species. A species should consist of strains with homologous DNAs. The degree of variation tolerated is not clearly defined and in arriving at a figure the limitations of the experimental techniques used must be considered. DNA/DNA hybridization has led to the reclassification of several species of lactic-acid bacteria. In some cases phenotypically different 'species' belong to the same genotype, while the reverse is also true and an apparently physiologically uniform species has been split. Species in which both types of change have been established are listed in Table 1. It is not always easy for microbiologists working outside taxonomy to understand and accept these changes. When physiologically different strains of commercial importance are united into a single geno-species it is probably necessary to define sub-species with the divisions of the original species. The clear separation of species which share the majority of properties normally measured is difficult but, in most cases, some simply determined properties have been found which are linked to the different genotypes. Exact species recognition could be important where the different species are filling ecological niches particularly suited to their requirements, but not to the other physiologically similar species (Garvie *et al.*, 1984).

2.2. Components of Cell Walls and Membranes

Kandler (1982) has shown the importance of cell wall structures in relation to bacterial evolution. The wall is important in free-living

TABLE 1

Changes in Classification of Lactic-acid Bacteria due to DNA/DNA Hybridization experiments

Species	% hybridization of DNA	Reference	Supporting data
Species no longer separated			
1. *Streptococcus lactis* (wild and starter strains) 2. *Streptococcus lactis* subsp. *diacetylactis* 3. *Streptococcus cremoris*	66–100	Garvie *et al.* (1981)	Lactate dehydrogenase General metabolic behaviour
1. *Leuconostoc mesenteroides*[a] 2. *Leuconostoc dextranicum* 3. *Leuconostoc cremoris*	78–100	Garvie (1976) Hontebeyrie & Gasser (1977)	LDH Glucose 6-phosphate dehydrogenase
1. *Lactobacillus lactis* 2. *Lactobacillus bulgaricus*	86	Simonds *et al.* (1971)	Various, see Buchanan & Gibbons (1974), p. 581
1. *Lactobacillus helveticus* 2. *Lactobacillus jugurt*	80–100	Dellaglio *et al.* (1974)	Various, see Buchanan & Gibbons (1974), p. 582
1. *Lactobacillus fermentum* 2. *Lactobacillus cellobiosus*	70–100	Buchanan & Gibbons (1974), p. 587 Vescovo *et al.* (1979)	Cell wall murein LDH
Species divided			
'Pediococcus cerevisiae' (plant origin) → *pentosaccus* / *acidilactici*	20–30	Dellaglio *et al.* (1981) Back & Stackebrandt (1978)	Some physiological properties LDH
Lactobacillus acidophilus → *acidophilus* / *gasseri* / various other groups	<30	Johnson *et al.* (1980) Lauer *et al.* (1980) Lauer & Kandler (1980)	LDH Cell wall peptide Glycero-3-phosphate dehydrogenase
Lactobacillus fermentum → *fermentum* / *reuteri*	<30	Kandler *et al.* (1980)	LDH, % G + C Cell wall murein Some physiological properties

[a] Some strains have only about 50% DNA hybridization with the type strain and may be a separate species.

organisms and if its structure developed early in evolution then it is logical to use it in taxonomy.

2.2.1. Peptidoglycans of Cell Walls

Peptidoglycan types have been extensively studied in Eubacteria (Schleifer & Kandler, 1972). Most lactobacilli are of the L-Lys–D-Asp type, the exceptions being *Lactobacillus plantarum* (mDAP direct) and *Lb. viridescens* and *Lb. confusus* (L-Lys–L-Ala, L-Ser with variations), a pattern seen also in the leuconostocs—this relationship will be discussed again later. The streptococci have a variety of patterns suggesting a heterogeneous genus but the groupings do not coincide with those from rRNA/DNA hybridization studies (see Table 4, p. 51).

Cell wall sugars, together with the presence of mycolic acid (see Table 8, p. 62), are used to separate genera in the 'coryneform group' of bacteria. A rogue strain in any pattern would be immediately suspect, and further study required to establish its true identity. Cell wall studies are valuable because they can be used with any bacteria while other cell constituents will occur only when certain metabolic pathways are present.

2.2.2. Menaquinones of Cell Membranes

Menaquinones have been studied in a variety of bacteria but they are of limited taxonomic use with species important in cheese and dairy products. The lactic-acid bacteria are fermentative and lack cytochromes and the associated electron transport systems. Menaquinones have been detected in some enterococci and lactic streptococci (Collins & Jones, 1979) but not in other species. They are also present in propionibacteria (Swartz, 1973). In the streptococci the same pattern is found in *Streptococcus lactis, Str. lactis* subsp. *diacetylactis* and *Str. cremoris* and this is different from that in *Str. faecium* and *Str. faecalis* (Collins & Jones, 1979).

2.3. Metabolic Pathways and Enzymes

Bacteria have been described as bags of enzymes and it is this aspect which is of prime consideration in the manufacture of cheese and fermented dairy products. However, these enzymes, and the metabolic pathways whose reactions they catalyse, are also fundamental in bacterial taxonomy. It is logical to expect that the basic metabolism of closely related species would be the same but the enzymes involved can show major differences between species of any bacterial genus. These differences have been used successfully in taxonomy. The lactic-acid bacteria and their important enzymes have been the interest of a number of studies, while other genera important in cheese manufacture have not been examined to the same extent.

The lactic-acid bacteria can be divided into two groups depending whether their main pathway of glucose fermentation is the Embden–Myerhof (EM) glycolytic pathway (homofermentative) or a combination of the hexose monophosphate (HMP) followed by the phosphoketolase pathways (heterofermentative) (Fig. 1, Table 2). However, some species normally forming only lactic acid also have enzymes associated with heterofermentative pathways. The pattern of glucose fermentation suggests that the betabacteria might have closer links with the leuconostocs than with the homofermentative lactobacilli; Fig. 1 shows the breakdown of glucose by both types of lactic-acid bacteria, the key enzymes in both pathways, and the end products formed. When substrates other than glucose are fermented or glucose is fermented in conditions of stress, metabolism can be affected and other end products may be formed.

Understanding the metabolic pathways clarifies the taxonomy of bacteria and also helps to explain their importance in fermentation (see Chapters 3, 6 and 7). In the lactic-acid bacteria the pathway of glucose fermentation can be ascertained by the end products formed: CO_2, acetate, ethanol and lactic acid. The fermentation route can also be established by detecting key enzymes, but as these are intracellular, broken cell extracts are used. The presence of fructose-1,6-diphosphate (FDP) aldolase shows that strains have the EM pathway while glucose-6-phosphate dehydrogenase (G-6-PDH) indicates heterofermentative species. The other key enzyme xyulose-5-phosphate phosphoketolase is less easily demonstrated because the direct substrate is unavailable commercially. The end products of fermentation, particularly CO_2 and lactic acid, have been used in the taxonomy of lactic-acid bacteria for a long time. The study of enzymes is recent and helps to separate species with the same fermentation pathways.

Lactate dehydrogenases (*LDH*). These enzymes are found in all lactic-acid bacteria (Garvie, 1980), and catalyse the final step in their energy-producing metabolism. There are three types of NAD dependent LDH, forming: (a) D($-$) lactate, (b) L($+$) lactate, (c) L($+$) lactate but requiring activation by FDP. The last type is found in streptococci, with the exception of *Str. thermophilus,* and in *Lb. casei.*

Ellen I. Garvie

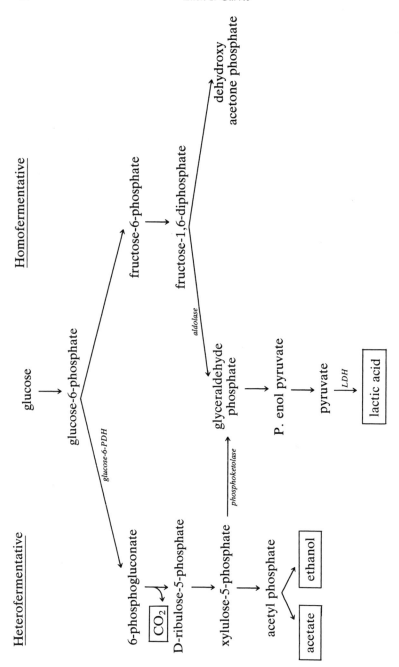

Fig. 1. Glycolytic pathways in lactic-acid bacteria.

TABLE 2
Metabolic Pathways in Lactic-acid Bacteria

| Genus | Sub-genus | Metabolic pathway | | |
		Embden–Myerhof	Hexose-monophosphate	Phospho-ketolase
Streptococcus	All species	+	−	−
Leuconostoc	All species	−	+	+
Pediococcus	All species	+(probably)	−	−
Lactobacillus	Thermobacteria	+	−	−
	Streptobacteria	+	(+)	(+)
	Betabacteria	−	+	+

+, present; (+), present but used only under special conditions; −, not present.

Table 3 gives the type of lactate formed by lactic-acid bacteria important in dairying. Electrophoresis of crude cell extracts separates the LDHs of different species (Garvie, 1969; Gasser, 1970; Back, 1978). From these studies it was shown that the LDHs of the non-acidophilic leuconostocs *Lb. confusus* and *Lb. viridescens* were electrophoretically identical, giving support to the suggestions that these species are close together in evolution despite morphological differences which are often

TABLE 3
Isomer of Lactate Formed by Dairy Lactic-acid Bacteria

D(−)	L(+)	DL
All leuconostocs	All streptococci	Lactobacillus $\begin{cases} helveticus \\ plantarum \end{cases}$
Lactobacillus $\begin{cases} lactis \\ bulgaricus \end{cases}$	Lactobacillus casei (a trace of D(−) is also found)	Lactobacillus $\begin{cases} fermentum \\ brevis \\ buchneri \\ confusus \\ viridescens \end{cases}$
		Pediococcus $\begin{cases} pentosaccus \\ acidilactici \end{cases}$

difficult to see. Most LDHs can be detected in gels using the appropriate lactate, though some lactobacilli have L(+) LDHs which are virtually irreversible and react only with pyruvate. This is true also of LDHs activated by FDP and location of enzyme bands of these enzymes after electrophoresis is not a satisfactory taxonomic technique. These FDP activated enzymes are inhibited by phosphate and the effect on enzyme activity of both FDP and phosphate is the same for enzymes of any species but different for the LDHs of different species (Garvie et al., 1981). The affinity for pyruvate at different pH values also differs between LDH of various species.

Some properties of the LDHs of *Str. lactis* and its sub-species and of *Str. raffinolactis* have been published (Garvie, 1978; Garvie et al., 1981). The LDH of *Str. faecalis* is unusual because phosphate stimulates its activity at pH 7·0 but slightly inhibits at pH 5·5. It resembles the LDH of *Str. raffinolactis* in having a low affinity for pyruvate at pH 7·0 and markedly increased activity at pH 5·5 compared with that at pH 7·0. The LDH of *Str. faecium* also has a low affinity for pyruvate at pH 7·0. Phosphate has only a slight effect on this enzyme which has a wide tolerance of pH. Unlike LDHs of other streptococci, that of *Str. thermophilus* is unaffected by FDP.

Additional separation of species is achieved by using immunological studies of enzyme proteins and this has been used for the LDHs of some lactic-acid bacteria (Gasser & Gasser, 1971; Hontebeyrie & Gasser, 1975). Studies on LDHs indicated that certain species were heterologous and, conversely, that different species, having the same LDH, were only sub-species. DNA/DNA hybridization studies have subsequently confirmed these findings (see Table 1). If immunological studies were applied to streptococcal LDHs the general classification of the genus might be clarified.

Glycolytic enzymes. Taxonomic use has been made of glucose-6-phosphate dehydrogenase (G-6-PDH) in heterofermentative organisms (Hontebeyrie & Gasser, 1975) and fructose-1,6-diphosphate (FDP) aldolase in homofermentative organisms (London & Kline, 1973). Phylogenetic maps prepared using information from LDHs (Gasser & Gasser, 1971) and aldolase (London & Chace, 1976) show the same relationships between species of lactobacilli. The findings from studies of aldolase showed that *Str. faecalis* and *Lb. casei* were close while *Pediococcus pentosaccus* and *Pediococcus acidi-lactici* were closer to the streptobacteria (sub genus of the lactobacilli) than to the streptococci. Recent work using glyceraldehyde-3-phosphate dehydrogenase (London & Chace, 1983) has given results in general agreement with those of earlier studies using different enzymes as markers. There is therefore increasing evidence from these biochemical studies that the morphologically-based separation of lactic-acid bacteria into two families is unsound.

2.4. Physiological Tests

For over half a century bacteria were classified largely on their ability to carry out certain chemical changes. These varied from genus to genus and their selection was largely empirical. The vitamin and amino acid requirements of bacteria were also studied and used in taxonomy but precautions necessary to ensure reproducible results exclude these properties as aids to routine examination. Complete agreement between all strains on all tests is an ideal never reached, the variability causes problems and also gives taxonomy the reputation of being a difficult and unreliable branch of microbiology. Much identification still relies on physiological tests with the consequent problems of recognizing every strain examined. In using physiological tests, only a small fraction of the genome can be used. The reason why properties are suppressed is not understood but among important diary species there is evidence that a large part of the genome has been suppressed as species have adapted to growth in a milk based environment. Where this has happened to commercially important bacteria sub-species are created. Thus *Leuconostoc mesenteroides* has three sub-species *mesenteroides, dextranicum* and *cremoris* (Garvie, 1983) and *Str. lactis* also has three sub-species *lactis, diacetylactis* and *cremoris* (Garvie & Farrow, 1982). In the case of *Str. lactis* it may be desirable to create a fourth sub-species to include the wild strains whose pattern of lactose hydrolysing enzymes (β-galactosidases) differs from that of starter strains (Farrow, 1980). Physiological tests in use do not separate wild and starter strains. Details of physiological tests for the identification of important dairy species will be given later.

Miniaturized kit systems are available for the determination of phenotypic properties and computers used for the interpretion of the results. 'Kit systems' were developed and are useful mostly for Gram-negative bacteria. Identification of dairy bacteria is not satisfactory with these mass produced testing kits, which have not been designed specifically for them (Griffiths & Phillips, 1982). The properties assessed, while numerous, may not include those particularly relevant to a fermentation process. In genera like the streptococci, which include a wide range of species from a variety of habitats, a general range of phenotypic properties is not particularly useful. This is one genus where careful selection of a few tests can partially identify strains; a more elaborate study, again using selected tests, will give a more exact identification.

Interpretation of a large mass of information about phenotypic properties of many strains is time consuming and computer analysis of such data has been used for many years. This approach is attractive when facilities are available. Many properties, however, including the information on nucleic acid and proteins, do not lend themselves to numerical analysis because it is difficult to arrange the information in a suitable form. Morphology poses the same problems. Numerical analysis will satisfactorily arrange strains into groups on the basis of similarity and give a quantitative estimation of the degree of similarity. Under a variety of situations this is useful and strains will be identified. The question must be asked whether numerical taxonomy is true taxonomy and not just a method of grouping like with like. There are several genera in which numerical analysis of phenotypic properties has arranged strains in groups which correspond to species based on information from nucleic acid studies.

This situation does not apply universally and may be truer with Gram-negative than with Gram-positive species. When the results of Bridge & Sneath (1983) using numerical taxonomy are compared with the results from studies using nucleic acids it will be seen that the two approaches show some differences in the relationships between species and genera. Only nucleic acid studies indicate that the present genus *Streptococcus* should be divided into several genera.The lactic-acid bacteria were studied and classified using phenotypic properties (e.g. Orla-Jensen, 1919) before the advent of the computer. In many instances this classification has remained unchanged with increased information from studies of cell components including nucleic acids. There remains the problem of the species listed in Table 1 where the phenotypic properties selected separate genetically homogeneous units on the one hand but fail to separate genetically diverse units into their different groups. In these cases phenotypic classification, whether computer assisted or not, has proved unreliable in recognizing species. There are arguments, therefore, that classification based only on phenotypic properties is a secondary technique to be applied after the genetic species are known. Phenotypic properties separating the units required by the investigator can then be selected. Several recent papers have discussed the value of the numerical approach to taxonomy (Harvey & Pickett, 1980; Kandler & Schleifer, 1980).

2.5. Morphological Tests

Even before it was possible to culture bacteria they were observed and, naturally, grouped according to form. This concept is basic to the classification scheme in use at present, but is it sound? Fox *et al.* (1980) criticize the heavy emphasis traditionally placed on morphology and consider spherical shape to be the principal offender. Several reasons have already been mentioned for considering the grouping of leuconostocs and pediococci with streptococci to be misjudged. Leuconostocs and lactobacilli can be difficult to separate morphologically (Garvie, 1975), and cells of *Str. lactis* may be coccoid when cultured in milk but are often elliptical in broth culture.

The morphology of freshly isolated wild strains is usually determined early in any study, and misinterpretation can lead to problems in identification. Should the traditional emphasis on morphological evidence be reduced, so that all the lactic-acid bacteria could be placed in a single family?

Even using modern scientific evidence it is difficult to gain acceptance for fundamental changes in classification. Perhaps now that a firmer basis for taxonomy is developing, corrections will be more readily accepted than in the past. Only a few key enzymes have been studied and the nature of the work limits the numbers of strains which can be examined. However, it is on the evidence from nucleic acid and enzyme studies that the plan of bacterial evolution will be based. This type of work is not suited to the rapid identification of large numbers of strains but the aim must be to find methods which can be used for large numbers of isolates which will give an answer showing true evolutionary species.

Bacterial plasmids and other extra chromosomal elements (ECE) which can move from cell to cell have been used in arguments supporting the theory that a true bacterial species does not exist. The movement of ECE in wild bacterial populations is difficult to establish, but bacterial species are remarkably stable (Woese, 1982; Reanney *et al.*, 1983). Plasmids will affect the physiological properties of a strain, when the plasmid's presence is expressed in changed metabolism. In time, the role of plasmids in evolution will become clearer and, as this happens, phylogenetic maps may have to be modified. Taxonomy is still a fluid science and anyone trying to identify bacteria must be able to accept the changes as they become necessary.

The metabolism of other genera of dairy bacteria have not been studied to the same extent as that of the lactic-acid bacteria. Brevibacteria and propionibacteria are both important in cheese making but their enzymes have not yet been used in their classification.

3. THE SIGNIFICANCE OF MILK IN THE EVOLUTION OF LACTIC-ACID BACTERIA

The most important lactic-acid bacteria in commercial use were isolated from cultures used in making traditional products, and are not known to exist outside the dairy environment. It would seem, therefore, that milk has influenced the evolutionary development of these organisms. Lactose is the only β-galactoside to occur naturally and it is found only in mammalian milk. Although bacteria in the intestinal tract of young (suckling) animals will encounter lactose, amounts will be limited by the activity of the β-galactosidase (β-gal) of the host. Lactose would also be available to bacteria living in or on the mammary gland. However, intestinal and skin bacteria are not used in fermented milk products and those bacteria which are important in their preparation are not normally associated with animals.

Str. lactis subsp. *cremoris,* of prime commercial importance today in cheese starter cultures, has a restricted fermentation pattern and a habitat confined to laboratories and environments in contact with dairies. *Str. lactis* subsp. *lactis* is also used as a cheese starter. It can use a wider range of carbohydrates than *Str. cremoris* but starter strains of sub-species *lactis* are restricted to the same habitats as sub-species *cremoris.* Wild type *Str. lactis* are found on plants but it has long been known that these ferment milk slowly and they do not form a reserve of possible starter strains. Recently it was found that wild *Str. lactis* have a low level of lactose hydrolysing enzymes. β-gal may be present and also β-phospho-galactosidase (β-P-gal). Only the latter enzyme is found in starter strains (Farrow, 1980). Thus within a single genospecies (Garvie *et al.,* 1981; Jarvis & Jarvis, 1981) there is a different lactose hydrolysing system and a different metabolic capability. When, how and why did these changes occur and is it possible to envisage a time when the DNA of *Str. lactis* subsp. *cremoris* could change sufficiently to be classed as a separate species? Starter streptococci have a battery of plasmids and some of these are concerned with lactose fermentation. Should it transpire that the variations between the sub-species of *Str. lactis* are due to plasmids then it would show that plasmids, while affecting phenotypic properties, have little impact on geno-species. The evolution and development of the different phenotypic components of *Str. lactis* are intriguing.

A similar range of phenotypic variants are found in *Leuc. mesenteroides* with its sub-species *mesenteroides dextranicum* and *cremoris* (Garvie, 1983). Here again the sub-species in which there has been a major suppression of phenotypic properties is the one most closely associated with milk and starters, i.e. *cremoris*. A further example is found among the lactobacilli where marked suppression of phenotypic properties may have been involved in the evolution of *Lb. bulgaricus* (dairy strains) from *Lb. lactis* (wild strains) (Buchanan & Gibbons, 1974).

Str. thermophilus is also highly adapted to live in milk. It ferments few sugars, among which disaccharides such as lactose are commonly preferred to glucose, and is metabolically inactive in many other respects. *Str. thermophilus* is found in milk and yoghurt and seldom elsewhere. Recent work (Kilpper-Bälz *et al.*, 1982; Farrow & Collins, 1984) has shown high homology between the DNA of *Str. thermophilus* and that of *Str. salivarius*. The significance of the findings is discussed later.

4. IDENTIFICATION OF BACTERIA IMPORTANT IN CHEESE AND FERMENTED MILKS

4.1. Lactic-acid Bacteria
4.1.1. Streptococci
Different species of streptococci occupy different habitats and the taxonomy of the genus has developed in separate pockets defined, virtually, by habitat, i.e. human or animal pathogens, dental flora (mainly viridans streptococci), enterococci, lactic streptococci etc. A recent attempt to survey the whole genus was made by Bridge & Sneath (1983). The present discussion will deal only with starter streptococci and other species which will grow in milk and might occur along with starter organisms, i.e. *Str. raffinolactis, Str. faecalis* and *Str. faecium.* Streptococci associated with bovine mastitis are not included although they also occur in milk; they are destroyed by pasteurization

or usually overgrown by other species. *Str. bovis* may also be found in milk but requires further study as there is considerable variation in the properties of different strains (Garvie & Bramley, 1979).

Serology has played a considerable part in classifying streptococci but usually into broad groups rather than into species. This approach is useful for pathogens but of doubtful value for non-pathogens. Serology has been of little help in the taxonomy of the viridans streptococci (Hardie & Bowden, 1976) and is misleading with Group D streptococci (Farrow *et al.*, 1983). Serology is probably best forgotten when working with dairy streptococci.

Mesophilic species. Techniques for studying lactic streptococci are described in the literature and references are given with Table 4 which gives the properties of these species. When only their cultural and physiological properties were known it was reasonable to divide the lactic streptococci into 3 species. Eventually, however, *Str. diacetylactis* was recognized as a sub-species of *Str. lactis* (Buchanan & Gibbons, 1974). The only essential difference between the two is the ability of *Str. diacetylactis* to metabolize citrate to acetoin/diacetyl and CO_2; other strains of *Str. lactis* do not utilize this substrate. Diacetyl is important in the flavour of dairy products but may be overproduced by some strains of streptococci, while the CO_2 formed can cause faults (see Chapters 6, 7). For these reasons the Dairy Industry needed to distinguish the citrate utilizing strains. Citrate uptake is known to be plasmid encoded (see Chapter 4) and is therefore an unreliable property on which to separate species.

When the growth and fermentation properties of *Str. lactis* and *Str. cremoris* are compared, reasons for merging these two species are less obvious. However, their biochemical similarities are now becoming recognized and Lawrence *et al.* (1976) were the first to suggest that all starter streptococci belonged to a single species. The LDHs of all strains of starter streptococci and wild strains of *Str. lactis* studied have the same properties and this enzyme is distinct from that in all other species. The ability to hydrolyse arginine has been used as a key property in separating *Str. lactis* and *Str. cremoris* (it is rather variable in *Str. diacetylactis*). A recent study of arginine metabolism has shown that separation using this property is only relative (Crow & Thomas, 1982) and these authors conclude that arginine metabolism gives support to the concept of a single species covering *Str. lactis* and *Str. cremoris*.

The evidence from DNA/DNA hybridization studies gives additional

TABLE 4
Properties of Streptococci Important in Cheese and Fermented Milks and of Other Species Found in the Same Environment

	Streptococcus lactis subsp.			Streptococcus raffinolactis	Streptococcus faecalis	Streptococcus faecium	Streptococcus thermophilus
	lactis	diacetylactis	cremoris				
[a] Growth at 10°C	+	+	+	+	+	+	−
40°C	+	+	−	−	+	+	+
45°C	−	−	−	−	+	+	+
[a] Growth in 2% NaCl	+	+	+	+	+	+	±
4% NaCl	+	+	−	−	+	+	−
6·5% NaCl	−	−	−	−	+	+	−
[a] Survival after 60°C for 30 min	−	−	−	−	+	+	+
[a] Hydrolysis of arginine	+	±	−	−	+	+	−
[a] Acid formed from							
arabinose	−	−	−	±	−	±	−
ribose	±	u	u	+	u	u	−
xylose	±	−	−	±	u	u	−
sorbose	−	−	−	−	u	u	−
lactose	+	+	+	+	+	+	+
maltose	+	+	−	+	+	+	−
melizitose	−	−	−	u	u	−	−
sucrose	±	±	−	+	+	+	+
trehalose	±	+	−	+	+	±	−
raffinose	±	+	−	+	+	+	−
dextrin	−	+	−	±	−	−	+
mannitol	±	±	−	±	u	±	−
sorbitol	−	−	−	+	+	±	+
aesculin	+	+ or slight	−	+	+	+	−
salicin	+	+	−	+	+	+	−
[b] % G + C in DNA	34·5–36[b]	35·2[b]	34·5–35·5[b]	40·5–42·5[b]	37–39[c]	37–38[c]	39–40
[d] RNA homology group	lactis	lactis	lactis	u	faecalis	faecalis	mastitis
[e] Peptidoglycan	←——— L-Lys-D-Asp ———→			L-Lys-L-Ala	L-Lys-L-Ala	L-Lys-D-Asp	L-Lys-L-Ala

+, over 90% strains +ve; −, less than 10% strains +ve; ±, between 10 and 90% strains +ve; u, unknown.
[a] For methods see Swartling (1951); [b] figures from Garvie et al. (1981); [c] figures from Roop et al. (1974); [d] see Garvie & Farrow (1981); [e] see Schleifer & Kandler (1972).

support (Garvie *et al.,* 1981; Jarvis & Jarvis, 1981). DNA/RNA hybridization studies show that *Str. lactis* and its sub-species belong to a single unit so far unassociated with any other species of streptococcus (Garvie & Farrow, 1981; Kilpper-Bälz *et al.,* 1982). The formation of two sub-species, viz. *lactis* and *cremoris,* is justified by the large phenotypic difference between the sub-species and by their different importance as starter cultures in the dairy industry.

Str. raffinolactis is the least known of the species included in Table 4. Reported isolations are few but it is possible that *Str. raffinolactis,* if isolated on other occasions, was not differentiated from the better known lactic-streptococci. Other streptococci are known to have phenotypic properties which are like those of *Str. lactis* (Garvie *et al.,* 1981; Collins *et al.,* 1983). These bacteria are found in mastitic milk and could well be mistaken for *Str. lactis.* They have a higher % G + C (38–39) than *Str. lactis,* a different LDH, and are separated by DNA/DNA hybridization from other streptococci. Thus although streptococci have been extensively studied for nearly 100 years new species are still being found. The dairy industry requires a clear identification of those species important to it.

Lactose hydrolysing enzymes β-gal and β-P-gal are of major importance in dairy streptococci. These enzymes have not been extensively studied but some variation in pattern is appearing. The significance to taxonomy may be slight because lactose breakdown is plasmid borne (see Chapter 4). Within *Str. lactis* and its sub-species considerable variation is found (Farrow, 1980). Wild strains may have little β-gal and even less β-P-gal, while starter strains have a high level of β-P-gal and no β-gal. The presence of β-P-gal indicates the presence of the tagatose pathway (see Chapter 3) but whether this is present in wild strains of *Str. lactis* is not known; it is unlikely to be present in species which have only β-gal. A high level of β-P-gal is not always associated with rapid clotting of milk. Both *Str. raffinolactis* and *Str. faecalis* have β-P-gal but neither clots milk rapidly. *Str. faecium* has both galactosidases (Farrow, 1980). A limited number of strains and species have been examined and more information is needed before the pattern of lactose hydrolysing enzymes and the subsequent fate of the products of hydrolysis in different species of streptococci are understood. β-P-gal and certain components of the lactose uptake system are plasmid encoded in some lactic streptococci (see Chapter 4). It is not yet certain, however, how widely the lactose hydrolysing enzymes are coded only on plasmids in different species.

The dairy industry is concerned with strains which ferment milk rapidly and this appears to be a property possessed only by adapted strains of *Str. lactis*. Strains of other species will lower the pH of milk to coagulate protein but at a slower rate. It should be noted that β-gal can be estimated after bacterial cells have been treated to make them permeable but only extracts of broken cells can be used for reliably estimating β-P-gal (Farrow, 1980).

Thermophilic species. Str. thermophilus has several properties distinguishing it from other streptococci. The enzymes at both ends of the glycolytic pathway are different from those of cheese starter streptococci. The LDH, unaffected by FDP, has been mentioned previously and lactose is hydrolysed by β-gal. The glucose half of the lactose molecule is used while the galactose is rejected. High DNA homology was found between two strains of *Str. thermophilus* and two of *Str. salivarius* (Kilpper-Bälz *et al.*, 1982). This relationship has been confirmed in a more extensive study by Farrow & Collins (1984) who propose that *Str. thermophilus* should be *Str. salivarius* subsp. *thermophilus* although they are worried by the large phenotypic differences between the organisms. The change in status of other species of lactic-acid bacteria has also been based on DNA/DNA homology but with the backing of enzyme and RNA studies. The LDHs of the type strains of *Str. thermophilus* and *Str. salivarius* have different properties (Garvie, unpublished) and LDH type and geno-species are generally in agreement. Garvie & Farrow (1981) put *Str. thermophilus* in a different RNA group to *Str. salivarius* (although these may only be subgroups). The RNA groups found by Garvie & Farrow and by Kilpper-Bälz *et al.* show some discrepancies and there is a different relationship of *Str. thermophilus* and *Str. salivarius* to other streptococcal species in the two studies. Clarification is needed and meantime the relationships and ecology of *Str. thermophilus* remain uncertain. Biochemical and enzyme studies would help to resolve the position.

4.1.2. Leuconostocs

Leuconostocs are found in starter cultures and may be important in flavour formation because they break down citrate, forming diacetyl from the pyruvate produced (see Chapter 7). The pathway of utilization is the same in all lactic-acid bacteria able to use citrate. The initial break gives oxaloacetate and acetate. Oxaloacetate releases CO_2 and pyruvate is formed. When this pyruvate is in excess over that required for lactate production it is removed from the cell after conversion to other substances.

The leuconostocs are less active than *Str. lactis* sub-species *diacetylactis*, attacking citrate only in acid media, and moreover this property may be lost in laboratory strains. A few properties which will separate the different sub-species and species of leuconostoc are given in Table 5. More information on the taxonomy of the genus will be found in Garvie (1984).

Leuconostocs can be divided into two groups: (a) the acidophilic *Leuc. oenos* which does not occur in milk and milk products and (b) non-acidophilic species which comprise the three sub-species of *Leuc. mesenteroides*, i.e. *mesenteroides*, *dextranicum* and *cremoris*, *Leuc. paramesenteroides* and *Leuc. lactis*. The species listed under (b) belong to the same RNA type (Garvie, 1981) but can be separated by DNA/DNA homology (Garvie, 1976; Hontebeyrie & Gasser 1977), and by immunological studies of their LDHs and G-6-PDHs. Electrophoresis does not differentiate these enzymes from the different

TABLE 5

Phenotypic Properties Separating Species of Leuconostocs Found in Milk and Milk Products

	Leuconostoc mesenteroides subsp.			Leuconostoc para-mesenteroides	Leuconostoc lactis
	mesenteroides	dextranicum	cremoris		
Mucosid colonies on sucrose agar	+	+	−	−	−
Dissimilation of citrate	±	±	+	±	±
Fermentation of					
arabinose	+	−	−	±	−
xylose	±	±	−	±	−
fructose	+	+	−	+	+
galactose	+	±	$+^w$	+	+
lactose	$±^w$	$+^w$	$+^w$	$±^w$	+
maltose	+	+	−	+	+
melibiose	±	±	−	+	±
sucrose	+	+	−	+	+
trehalose	+	+	−	+	−
aesculin	±	±	−	±	−
salicin	±	±	−	−	±

+, more than 90% strains +ve; −, less than 10% strains +ve; ±, between 10 and 90% strains +ve; ^{w} weak reaction.

species. Phenotypic properties are quite variable and a clear diferentiation using these characters is problematical. Lactose is not readily utilized and information about lactose hydrolysing enzymes in leuconostocs is not available. *Leuc. lactis* can grow in milk and form acid but other species usually fail to form acid in milk.

Any leuconostoc strains used in the dairy should be screened for their ability to utilize citrate (at pH values below 5.0) as only strains able to do so are of use in starter cultures. Occasional checking is necessary because citrate utilization may be an unstable property.

Leuc. lactis can be isolated from raw milk but *Leuc. mesenteroides* sub-species *cremoris* is an adaption to milk and all strains known originated from creameries. Care should be taken in preserving these strains (if they are important) because replacements will be hard to find.

Leuconostocs can be separated from streptococci because they form gas from the fermentation of glucose and produce D(−) lactate. They can be separated from the gas forming lactobacilli which form DL lactate, however *Lb. viridescens* forms mainly D(−) with often less than 10% L(+) lactate but it is seldom found in milk.

4.1.3. Lactobacilli

There are fundamental differences between species at present classed as lactobacilli with morphology as the linking property. Most species of lactobacilli share habitats and the genus has been studied as a whole for a long time by many investigators without serious consideration as to whether it was a logical classification. Arguments can be used to justify splitting the genus into those with a primary EM pathway and those in which this fermentation is absent. The latter then link with the leuconostocs. In dairying, homofermentative lactobacilli are used in combination with *Str. thermophilus* and these have been naturally selected because the temperatures used in certain cheese making processes (and in yoghurt) are too high for group N streptococci. The lactobacilli with lower growth temperatures (*Lb. casei* and *Lb. plantarum*) occur in many cheeses but they are natural contaminants. The starter lactobacilli are saprophytes and absent from the intestinal tract of animals, but the species found in animals may be natural contaminants of dairy products. In the lactobacilli, as in other genera, ecology is a useful guide to species identification.

Lb. lactis and *Lb. helveticus* are the main starter lactobacilli for Swiss type and other cheese with high make temperatures, while *Lb.*

Ellen I. Garvie

bulgaricus is a component of yoghurt starter and a heterofermentative lactobacillus important in kefir. The bacteria and yeasts of kefir have not been as well studied as the flora of yoghurt, probably because commercial production is less important. Conditions of manufacture of kefir are not as well controlled as they are for other diary products, with a resulting possible variation in microbial flora. A new species *Lb. kefir* is described by Kandler & Kunath (1983) and this species has been found in kefir in the UK (Marshall *et al.*, 1984). The phenotypic properties of these species are given in Table 6.

The classification of lactobacilli is given in Buchanan & Gibbons (1974) and also by Sharpe (1981).

Very little is known about the lactose hydrolysing enzymes in

TABLE 6

Differential Properties of Species of Lactobacilli Important in Milk and Dairy Products

Property	Lb. helveticus	Lb. lactis	Lb. bulgaricus	Lb. casei	Lb. plantarum	Lb. brevis	Lb. kefir
CO_2 from glucose fermentation	−	−	−	−	−	+	+
Ammonia formed from arginine	−	−	−	−	−	+	+
Type of lactic acid formed	DL	D(−)	D(−)	L + (a trace of D(−))	DL	DL	DL
Growth at 15°C	−	−	−	+	+	+	+
Growth at 45°C	+	+	+	±	−	−	−
Acid from							
arabinose	−	−	−	−	±	+	+
galactose	+	+	+	+	+	−	±
cellobiose	−	−	−	+	+	−	−
lactose	+	+	+	+	+	−	±
maltose	+	+	−	+	+	+	±
melibiose	−	−	−	−	+	+	±
sucrose	−	+	−	±[w]	+	±	−
trehalose	±	+	−	±	+	−	−
melizitose	−	−	−	+	±	−	−
raffinose	−	−	−	−	+	w	−
mannitol	−	−	−	+	+	w	−
sorbitol	−	−	−	+	+	−	−
aesculin	−	±	−	+	+	±	u
salicin	−	+	−	+	+	−	−
amygdalin	−	−	−	+	+	−	u
% G + C in DNA	39–40	49–51	49–51	45–47	44–46	43–46	41·5

+, more than 90% strains +ve; −, less than 10% strains +ve; ±, between 10 and 90% strains +ve; u, unknown; [w] weak reaction.

lactobacilli. Premi *et al.* (1972) screened strains of a number of species and β-gal was the dominant enzyme in *Lb. helveticus, Lb. lactis* and *Lb. bulgaricus. Lb. casei* did not have β-gal but some β-P-gal activity was recorded and no galactosidase was found in *Lb. buchneri* which, in any case, does not ferment lactose. There are several implications from this pattern. Swiss type cheese may have a high galactose content because lactic-acid bacteria with β-gal use only the glucose half of lactose. *Lb. casei* with β-P-gal is like the streptococci rather than the other lactobacilli. London *et al.* (1971) found a relationship between *Lb. casei* and *Str. faecalis* using immunological studies of two forms of malic enzyme. It is also known that like streptococci *Lb. casei* has an FDP dependent LDH, and its aldolase shares immunological homology with that of group D streptococci (London & Kline, 1973). The full significance of these relationships is not clear but they are a further indication of unity amongst lactic-acid bacteria.

4.1.4. Pediococci
Pediococci are not used in any dairy starters, though they may grow in maturing cheese and ferment residual lactose over a long period. They can also occur in milk and will grow in media selective for lactobacilli. Morphologically pediococci are true cocci dividing in two planes (Gunther, 1959) and can be mistaken for micrococci. However, they do not possess catalase and are homofermentative, producing large amounts of lactic acid when fermenting glucose. Only two species, *Ped. pentosaccus* and *Ped. acidilactici* are found in dairy products, neither ferment lactose actively, and milk is a deficient medium for a number of strains of the former species which have a requirement for folinic acid. Recent work on the classification of pediococci of all species is found in Back (1978).

4.2. Bacteria of the Coryneform Group
Fermentations in the dairy industry are traditionally dependent on lactic streptococci or *Str. thermophilus* with certain species of lactobacilli also included in starters when high temperatures are used. Both propionibacteria and brevibacteria, however, occur in certain types of cheese and are important, if not essential, in ripening. The propionibacteria may be a component of some starter cultures. Both propionibacteria and brevibacteria are classified in the coryneform group. This group has been studied in a variety of laboratories but the diary species have received less attention than those from other sources.

4.2.1. Propionibacteria

Classically the genus *Propionibacterium* contained bacteria found mainly in those Swiss cheese varieties in which eye formation is important (Emmental and Gruyere). Recently some 'coryneforms' associated with skin have been transferred to the genus because they form large amounts of propionic acid from sugar fermentation. These anaerobic skin bacteria have a lower % G + C in their DNA than the classical species (56–64% against 65–68%) (Johnson & Cummins, 1972), and are not further discussed here.

The classical propionibacteria were divided into a variety of species using phenotypic properties but these have been grouped into 4 as the result of DNA/DNA homology studies (Johnson & Cummins, 1972). Virtually the same grouping of species resulted from an earlier study where classification was based on morphological and physiological features (Malik *et al.*, 1968). The species are *Propionibacterium jensenii, Pr. thoenii, Pr. acidi-propionici* and *Pr. freudenreichii. Pr. jensenii* includes strains previously classified as *Pr. jensenii, Pr. technicum, Pr. raffinosaceum, Pr. peterssonii, Pr. rubrum* and *Pr. zeae. Pr. thoenii* includes *Pr. thoenii* and *Pr. rubrum. Pr. acidi-propionici* includes *Pr. arabinosum, Pr. pentosaceum,* and *Pr. shermanii. Pr. freudenreichii* includes *Pr. shermanii, Pr. freudenreichii* and *Pr. arabinosum.* From the above it will be seen that some 'species' contained misidentified strains.

Propionibacteria are anaerobic although they will adapt to aerobic growth in broth cultures when handled in the laboratory. Catalase and cytochromes together with an electron transport system are present (Hettinger & Reinbold, 1972; Swartz, 1973) although the organisms are anaerobic. Menaquinones are present, confirming the presence of electron-transport systems (Swartz, 1973). Glucose is metabolized to pyruvate by the EM pathway but there is also evidence of HMP pathway enzymes (Hettinger & Reinbold, 1972). NAD linked LDH is absent and pyruvate is converted to acetate and propionate. The enzymes of this pathway were purified by Allen *et al.* (1964) and the overall reaction (see Fig. 2) is:

$$3 \text{ glucose} \rightarrow 2 \text{ acetate} + 4H_2O + 2CO_2 + 4 \text{ propionate}$$

The strain used in this work was classified as *Pr. shermanii:* a comparison between the enzymes of the different species might be useful to clarify the taxonomy of this genus.

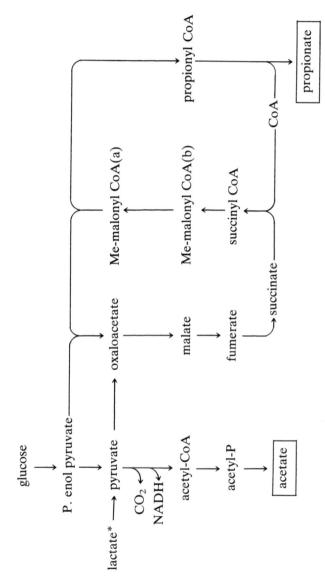

Fig. 2. Formation of propionate by *Propionibacterium shermanii* (after Allen *et al.*, 1964).

* In secondary cheese fermentation

TABLE 7

Differential Properties of Species of Propionibacteria found in Cheese

Property	Pr. freudenreichii	Pr. theonii	Pr. acidipropionici	Pr. jensenii
Wall composition				
DAP	meso	L	L	L
galactose	+	+	±	±
glucose	−	+	+	+
mannose	+	trace	trace	trace
rhamnose	+	−	−	−
% G + C in DNA	65–67	66–67	66–68	66–68
Fermentation of				
lactose	±	±	+	±
trehalose	−	+	$+^w$	$+^w$
mannitol	−	−	+	+
aesculin	+	+	+	+
Hydrolysis of gelatin	partial	partial	partial	partial
Reduction of nitrate	±	−	+	−
% DNA/DNA hybridization[a]	Average figure for several strains against reference DNA			
Pr. freudenreichii	90	20	25	26
Pr. theonii	12	96	30	53
Pr. acidipropionici	8	35	87	38
Pr. jensenii	17	51	30	88

+, more than 90% strains +ve; −, less than 10% strains +ve; ±, between 10 and 90% strains +ve; [a] from Johnson & Cummins (1972); [w] weak reaction.

The ratio of acetate : propionate formed can vary from the expected ratio of 1 : 2 and may be as high as 1 : 5, because these bacteria have subsidiary metabolic pathways including an NAD independent LDH enabling them to convert lactate to pyruvate (Molinari & Lara, 1960). Cheese ripened by propionibacterium is therefore distinct in flavour from that ripened by lactic-acid bacteria and the CO_2 produced forms the 'eyes' of Swiss-type cheese (for further details see Chapter 7). Some differential properties of the 4 species of propionibacteria are given in Table 7.

4.2.2. Brevibacteria

Brevibacterium linens occurs on the surface of a variety of soft cheeses e.g. Camembert, Limburger and Stilton, and is well known in the dairy industry. Taxonomically its affiliation to other bacterial species has been uncertain but recently it was recognized as belonging to the 'coryneform' group with a possible relationship to the arthrobacters. The genus *Brevibacterium* is listed in the Approved List of Bacterial Species (Skerman *et al.* 1980) with 21 named species. This is undoubtedly a heterogeneous collection of bacteria of which some may belong to the same genus as *Br. linens.* Recently non-pigmented organisms forming methanethiol were isolated from skin and Cheddar cheese. These organisms were thought to contribute to Cheddar cheese flavour and they have been identified as brevibacteria (Sharpe *et al.* 1976, 1977). Crombach (1974) has studied *Br. linens,* orange-pigmented coryneform organisms from sea-fish and non-pigmented coryneforms from cheese. Table 8 gives some properties of these organisms and of related genera. The absence of mycolic acid and arabinose in the cell wall separates *Brevibacterium* species from corynebacteria with mesodiamino pimelic acid. DNA/DNA hybridization has shown the probable species groups within the brevibacteria (Crombach, 1974, 1978). The soil arthrobacters were only remotely related to the orange cheese coryneforms which were mostly highly related to *Br. linens.* The non-pigmented brevibacteria were only remotely related to the orange ones. Similarly the orange sea strains were a distinct group with little DNA in common with the orange cheese strains.

The metabolism of brevibacteria has not been studied to the same extent as that of the lactic-acid bacteria but they are proteolytic rather than fermentative. The main enzymes have not been used in the taxonomy of the genus.

5. CONCLUSIONS

Economic pressures are forcing research work along paths dictated by commercial interests, and the significance of taxonomy is not clear to those who hold the purse strings. Too often the taxonomist is regarded as working in a remote world quite unrelated to any practical requirements. Only the taxonomist thinks in terms of bacteria and will look at a diversity of strains from as wide an ecological field as possible. The interests of the taxonomist can only be limited by the genetic diversity of bacteria. For the more practical approach it is important that strains are correctly identified because, in giving a name, a vast amount of information is expressed. Giving the wrong name can give a false lead as to the metabolic processes which have taken place and are con-

TABLE 8

Properties of Brevibacteria Isolated from Cheese and some Related Bacteria

	Brevibacterium linens	Cheese isolates (Mulder)		Skin isolates	Arthrobacter	True coryneforms[a]	Cheese isolates (NIRD)
		1	2				
Cell wall sugar	Gal[b]	Gal	Gal/Arb[c]	Gal	Gal	Gal/Arb	Gal
Mycolic acid	–	–	–	–	–	+	–
Type of DAP	meso	–	meso	meso	often meso	meso	meso
% G + C in DNA	62·5–64	60–61	65–67	–	63–72	57–60	–
Vitamin requirements	wide	+	+	u	biotin only	complex	u
Require ⎰ organic N	range of	u	+	u	+	complex	u
⎱ glutamic acid	growth req.	u	+	u	–	complex	u
Methanethiol formed	–	–	–	+	–	–	+
Morphology rods→cocci	+	+	+	+	+	+	+
Pigment	orange	–	–	–	–	–	–
Opt. temp. of growth	22–30	30	22–30	30–37	22–30	30–37	30–37
Survival 60°30′	–	–	–	±	–	–	+
NaCl tolerated	10–12	5–10	5–12	15	u	10–15	12–15
Utilization of							
lactose	–	–	–	u	±	u	±
sucrose	–	+	–	u	+	u	+
glycerol	+	+	+	u	+	u	u
acetate	+	+	+	+	+	u	u
lactate	+	+	+	+	±	u	+
Hydrolysis of starch	–	+	–	u	±	–	–
Liquefaction of ⎰ gelatin	+	+	–	+	+	–	+
⎱ milk	–	+	–	+	+	u	+
Tyrosine decomposed	+	u	u	+	u	±	+
Extracellular DNase	+	u	u	+	u	–	+

+, more than 90% strains +ve; –, less than 10% strains +ve; ±, between 10 and 90% strains +ve; u, unknown.

[a] Only species containing meso-DAP are included. [b] Gal, galactose; [c] Arb, arabinose

tinuing. Misidentification can lead to errors in interpreting the significance of the organism. There can be major differences in the metabolism of different species within a genus so that species identification can be important while broad generic groupings often have little meaning.

Genetic engineering is a useful tool but to use it logically it is desirable to know as much about the potential in the metabolism of a species and in related species. It is here that the taxonomist is important. The geneticist will work with useful strains, the taxonomist can point to useful properties in commercially useless strains. *Str. lactis* is an example where the variability within the species has been extended by learning of wild strains with a different metabolism from the starter strains.

Taxonomy is no longer limited to morphology and a study of easily determined phenotypic properties. It is now based on the whole organism. If new processes are to be successful or if the age-old processes are to be successfully modified, then changes must be based on the understanding of the capabilities of the microorganisms involved and on those that will develop naturally in the product. This information is more likely to be available from the study of bacterial taxonomy than from work in any other branch of microbiology. It is not unknown for important biochemical studies to be published without a careful check on the identity of the bacterial strain. Errors can be misleading and once in print they are difficult to eradicate.

A full picture can only be obtained where every genus is studied over its complete range of species. Work should not be restricted to bacteria useful or detrimental to man. As knowledge builds up it will become apparent how bacteria have evolved, how stable the species are and at what rate new species are developing. This information can be of value to the geneticist who, in altering the capabilities of strains, needs to make stable modifications.

A detailed study of *Str. lactis* (wild and starter strains) and its sub-species *cremoris* might answer some of the questions on species stability. How and when did the change to high β-P-gal activity arise? Why has a milk environment led to a suppression of many metabolic activities in *Str. cremoris* and why is this sub-species not found outside the dairy environment?

REFERENCES

ALLEN, S. H. G., KELLERMEYER, R. W., STJERNHOLM, R. L. & WOOD, H. G. (1964) *J. Bacteriol.* **87**, 171.

BACK, W. (1978) *Brauwissenschaft* **31**, 237, 312, 336.
BACK, W. & STACKEBRANDT, E. (1978) *Archive fur Mikrobiologie* **118**, 78.
BUCHANAN R. E. & GIBBONS N. E., Eds (1974) *Bergey's Manual of Determinative Bacteriology*, 8th edn. The Williams and Wilkins Co., Baltimore.
BRIDGE, P. D. & SNEATH, P. H. A. (1983) *J. General Microbiol.* **129**, 565.
CROMBACH, W. H. J. (1974) *Antonie van Leeuwenhoek* **40**, 347.
CROMBACH, W. H. J. (1978) In *Coryneform Bacteria*, I. J. Bousfield & A. G. Callely, eds. Academic Press, London, p. 161.
COLLINS, M. D. & JONES, D. (1979) *J. General Microbiol.* **114**, 27.
COLLINS, M. D., FARROW, J. A. E., PHILLIPS, B. A. & KANDLER, O. (1983) *J. General Microbiol.* **129**, 3427.
CROW, V. L. & THOMAS, T. D. (1982) *J. Bacteriol.* **150**, 1024.
DELLAGLIO, F., VESCOVO, M. & PREMI, L. (1974) *Int. J. Systematic Bacteriol.* **24**, 235.
DELLAGLIO, F., TROVATELLI, L. D. & SARA, P. G. (1981) *Zentralblatt fur Bakteriologie, Microbiologie und Hygiene,* 1 Abt. Originale **C2**, 140.
FARROW, J. A. E. (1980) *J. Appl. Bacteriol.* **49**, 493.
FARROW, J. A. E. & COLLINS, M. D. (1984) *J. General Microbiol.* **130**, 357.
FARROW, J. A. E., JONES, D., PHILLIPS, B. A. & COLLINS, M. D. (1983) *J. General Microbiol.* **129**, 1423.
FOX, G. E. *et al.* (1980) *Science* **209**, 457.
GARVIE, E. I. (1969) *J. General Microbiol.* **58**, 85.
GARVIE, E. I. (1975) In *Lactic Acid Bacteria in Beverages and Foods*, J. G. Carr, C. V. Cutting & G. C. Whiting, Eds. Academic Press, London, p. 339.
GARVIE, E. I. (1976). *Int. J. Systematic Bacteriol.* **26**, 116.
GARVIE, E. I. (1978) *Int. J. Systematic Bacteriol.* **28**, 190.
GARVIE, E. I. (1980) *Microbiol. Rev.* **44**, 106.
GARVIE, E. I. (1981) *J. General Microbiol.* **127**, 209.
GARVIE, E. I. (1983) *Int. J. Systematic Bacteriol.* **33**, 118.
GARVIE, E. I. (1984) In *Methods in Microbiology*, Vol. 16, T. Bergen, Ed. In press.
GARVIE, E. I. & BRAMLEY, A. J. (1979) *J. Appl. Bacteriol.* **46**, 557.
GARVIE, E. I. & FARROW, J. A. E. (1981) *Zentralblatt fur Bakteriologie, Mikrobiologie und Hygiene,* 1 Abt. Originale **C2**, 299.
GARVIE, E. I. & FARROW, J. A. E. (1982) *Int. J. Systematic Bacteriol.* **32**, 453.
GARVIE, E. I., FARROW, J. A. E. & PHILLIPS, B. A. (1981). *Zentralblatt fur Bakteriologie, Mikrobiologie und Hygiene,* 1 Abt. Originale **C2, 151.**
GARVIE, E. I., COLE, C. B., FULLER, R. & HEWITT, D. (1984) *J. Appl. Bacteriol.* **56**, 237.
GASSER, F. (1970) *J. General Microbiol.* **62**, 223.
GASSER, F. & GASSER, C. (1971) *J. Bacteriol.* **106**, 113.
GILLIS, M. & DELEY, J. (1980) *Int. J. Systematic Bacteriol.* **30**, 7.
GILLESPIE, D. (1968) *Methods in Enzymology* **12B**, 641.
GRIFFITHS, M. W. & PHILLIPS, J. D. (1982) *J. Appl. Bacteriol.* **53**, 343.
GUNTHER, H. L. (1959) *Nature (London)* **183**, 903.
HARDIE, J. M. & BOWDEN, G. H. (1976) *J. Dental Res.* **55**, A166.

HARVEY, S. & PICKETT, M. J. (1980) *Int. J. Systematic Bacteriol.* **30**, 86.
HETTINGER, D. H. & REINBOLD, G. W. (1972) *J. Milk Food Technol.* **35**, 295, 359, 437.
HONTEBEYRIE, M. & GASSER, F. (1975) *Int. J. Systematic Bacteriol.* **25**, 1.
HONTEBEYRIE, M. & GASSER, F. (1977) *Int. J. Systematic Bacteriol.* **27**, 9.
JARVIS, A. W. & JARVIS, B. D. W. (1981) *Appl. Environ. Microbiol.* **41**, 77.
JOHNSON, J. L. & CUMMINS, C. S. (1972) *J. Bacteriol.* **109**, 1047.
JOHNSON, J. L., PHELPS, C. F., CUMMINS, C. S. LONDON, J. & GASSER, F. (1980) *Int. J. Systematic Bacteriol.* **30**, 53.
KANDLER, O. (1982) *Zentralblatt fur Bakteriologie, Mikrobiologie und Hygiene,* 1 Abt. Originale **C3**, 149.
KANDLER, O. & KUNATH, P. (1983) *Systematic Appl. Microbiol.* **4**, 286.
KANDLER, O. & SCHLEIFER, K. H. (1980) *Progress in Botany* **42**, 234.
KANDLER, O., STETTER, K. O. & KOHL, R. (1980) *Zentralblatt fur Bakteriologie, Mikrobiologie und Hygiene,* 1 Abt. Originale **C1**, 264.
KILPPER-BÄLZ, R., FISCHER, G. & SCHLEIFER, K. H. (1982) *Current Microbiol.* **7**, 245.
LAUER, E. & KANDLER, O. (1980) *Zentralblatt fur Bakteriologie, Mikrobiologie und Hygiene,* 1 Abt. Originale **C1**, 75.
LAUER, E., HELMUNG, C. & KANDLER, O. (1980) *Zentralblatt fur Bakteriologie, Mikrobiologie und Hygiene,* 1 Abt. Originale **C1**, 150.
LAWRENCE, R. C., THOMAS, T. D. & TERZAGHI, B. E. (1976) *J. Dairy Res.* **43**, 141.
LONDON, J. & CHACE, N. M. (1976) *Archive fur Mikrobiologie* **110**, 121.
LONDON, J. & CHACE, N. M. (1983) *Int. J. Systematic Bacteriol.* **33**, 723.
LONDON, J. & KLINE, K. (1973) *Bacteriol. Rev.* **37**, 453.
LONDON, J., MEYER, E. Y. & KULCZYK, S. R. (1971) *J. Bacteriol.* **108**, 196.
MALIK, A. C., REINBOLD, G. W. & VEDAMUTHU, E. R. (1968) *Canadian J. Microbiol.* **14**, 1185.
MARSHALL, V. M., COLE, W. M. and FARROW, J. A. E. (1984) *J. Appl. Bacteriol.* **56**, 503.
MOLINARI, R. & LARA, F. J-S. (1960) *Biochem. J.* **75**, 57.
MOORE, R. L. & McCARTHY, B. J. (1967) *J. Bacteriol.* **94**, 1066.
ORLA-JENSEN, S. (1919) *The Lactic Acid Bacteria.* Høst, Copenhagen, pp. 1–196.
PREMI, L., SANDINE, W. E. & ELLIKER, P. R. (1972) *Appl. Microbiol.* **24**, 51.
REANNEY, D. C., GOWLAND, P. C. & SLATER, H. S. (1983) *Soc. General. Microbiol. Symp: Microbes in their Natural Environments,* J. H. Slater, R. Whittenbury & J. W. T. Wimpenny, Eds. Camb. Univ. Press, London.
ROOP, D. R., MUNDT, J. O. & RIGGSBY, W. (1974) *Int. J. Systematic Bacteriol.* **24**, 330.
SCHLEIFER, H. H. & KANDLER, O. (1972) *Bacteriol. Rev.* **36**, 407.
SHARPE, M. E. (1981) In *The Prokaryotes,* M. P. Starr, H. Stolp, H. G. Trüper, A. Barlows & H. G. Schlegel. Springer-Verlag, Berlin–Heidelberg, p. 1653.
SHARPE, M. E., LAW, B. A. & PHILLIPS, B. A. (1976) *J. General Microbiol.* **94**, 430.

SHARPE, M. E., LAW, B. A. & PHILLIPS, B. A. (1977) *J. General Microbiol.* **101,** 345.

SIMONDS, J., HANSEN, P. A. & LAKSHMANAN, S. (1971) *J. Bacteriol.* **107,** 382.

SKERMAN, V. B. D., McGOWAN, V. & SNEATH, P. H. A. (1980) *Int. J. Systematic Bacteriol.* **30,** 225.

SWARTLING, P. (1951) *J. Dairy Res.* **18,** 256.

SWARTZ, A. C. (1973) *Archive fur Mikrobiologie* **91,** 273.

VESCOVO, M., DELLAGLIO, F., BOTTAZZI, V. & SARA, P. G. (1979) *Microbiologica* **2,** 317.

WOESE, C. R. (1982) *Zentralblatt fur Bakteriologie, Mikrobiologie und Hygiene,* 1 Abt. Originale **C3,** 1.

WOESE, C. R. & FOX, G. E. (1977a) *J. Molec. Evolution* **10,** 1.

WOESE, C. R. & FOX, G. E. (1977b) *Proc. Nat. Acad. Sci. USA* **74,** 5088.

Chapter 3

The Physiology and Growth of Dairy Lactic-acid Bacteria

VALERIE M. E. MARSHALL and BARRY A. LAW

National Institute for Research in Dairying, Shinfield, Reading, UK

1. INTRODUCTION

Food fermentations which have traditionally been carried out by spontaneous growth and action of micro-organisms in the special environment afforded by food, are today carefully controlled microbial processes for which selected cultures have been developed. These starter cultures have been formulated by using micro-organisms which impart the special desired characteristics to the fermented product.

Starter cultures used for production of fermented dairy products vary enormously and include cultures of bacteria, moulds and yeasts (Table 1). These cultures produce lactic acid, acetic acid, aroma and CO_2. They are important also because they possess lipolytic and proteolytic activities in varying degrees. All the characters contribute to the fermentation and ripening into a final product with good organoleptic qualities.

Most of the lactic-acid bacteria used in the dairy industry belong to the genera *Streptococcus*, *Leuconostoc* and *Lactobacillus*. The ecology of the lactic-acid streptococci has been reviewed by Sandine *et al.* (1972) and a review of their classification and identification forms the second chapter of this book. Attention has been focused on the need of industry for new isolates of streptococci as yet unused for dairy fermentations.

The diversity of both cheeses and fermented milks makes a discussion of starter physiology complex. Milk can be fermented into well over a thousand products (Sandine & Elliker, 1970), each one different as a consequence of the controlled manufacturing environment which

TABLE 1

Micro-organisms for Fermentation of Dairy Products

Product	Micro-organisms added
Cheese	
Parmesan, Romano	Mixture of *Lactobacillus bulgaricus* and *Streptococcus thermophilus*
Cheddar, Colby	*Str. lactis; Str. cremoris; Str. diacetylactis*
Swiss, Emmenthaler	Mixture of *Lb. bulgaricus* (or *Lb. lactis* or *Lb. helveticus*) and *Str. thermophilus* and *Propionibacterium shermanii*
Provolone	Mixture of heat-resistant *Lactobacillus* species and *Str. thermophilus*
Blue, Gorgonzola, Roquefort, Stilton	*Str. lactis* plus *Penicillium roqueforti*
Camembert	Lactic streptococci plus *Penicillium camemberti* (caseicolum)
Brick, Limburger	Mixture of *Str. thermophilus* and *Str. cremoris;* mixture of *Str. thermophilus* and *Lb. bulgaricus;* mixture of *Str. lactis* and *Str. thermophilus*
Muenster	Mixture of *Str. thermophilus* and *Lactobacillus* species
Gouda, Edam	*Str. lactis, Str. cremoris,* Leuconostocs
Mozzarella	Mixture of heat resistant *Lactobacillus* species and *Str. thermophilus*
Cottage cheese, cream cheese	*Str. lactis* or *Str. cremoris;* mixture of *Str. lactis* or *Str. cremoris* and *Str. diacetylactis* (or *Leuconostoc* species)
Fermented milks	
Bulgarian buttermilk	*Lb. bulgaricus*
Acidophilus milk	*Lb. acidophilus*
Buttermilk, sour cream	*Str. lactis;* mixture of *Str. cremoris* and *Leuc. citrovorum* (or *Lb. dextranicum)*
Yogurt	Mixture of *Lb. bulgaricus* and *Str. thermophilus*
Villi	*Str. lactis, Str. cremoris, Str. diacetylactis, Leuc. citrovorum* plus *Oospora* species (or *Geotrichium candidum)*
Laban	*Str. thermophilus, Leuc. lactis, Lb. acidophilus Kluyveromyces fragilis, Saccharomyces cerevisiae*
Kefir	'Lactic-acid bacteria and yeasts'
Koumiss	*Lb. bulgaricus* and a torula yeast
Miscellaneous	
Butter	Mixture of *Str. diacetylactis* and *Str. lactis*
Cream dressing for cottage cheese	*Str. lactis; Str. cremoris; Str. diacetylactis*

induces the organisms to impart different flavours to the finished product. The primary requirement of most cultures is the production of lactic acid. *Streptococcus lactis* and *Str. cremoris* are used for cheese fermentation because they possess limited metabolic diversity and are homolactic during batch growth in milk, forming only small amounts of secondary products such as acetate and diacetyl. They are also weakly proteolytic and lipolytic. Carbon dioxide production from metabolism of *Leuconostoc* species and *Propionibacterium shermanii* is desirable where 'eye' formation is required in the finished cheese (Emmental, Gruyère) or a refreshing sparkling beverage is desired. The conversion of all of the lactose to lactic acid is important for cheese that is to be ripened because residual lactose may permit growth of contaminating organisms possessing greater biochemical diversity, with resultant production of off-flavours and changes in flavour and texture of the product (Chapter 7).

2. NUTRITION OF DAIRY LACTIC-ACID BACTERIA

Although the lactic-acid bacteria are widely distributed in nature, their nutritional requirements are complex. Their simple enzyme complement makes them unable to synthesize many amino acids and vitamins. Such requirements dictate the natural habitat of these organisms; for instance, growth of lactic-acid bacteria is rarely observed in water, but often in milk, milk products and meat where there are rich carbon and nitrogen sources. Some growth requirements for the lactic-acid bacteria important to dairying are shown in Table 2. The fact that the dairy lactic-acid bacteria grow in milk indicates that it can supply them with all of their essential nutrients such as vitamins (Reiter & Oram, 1962), metals (Olson & Qutub, 1970) and nucleic acid bases (Taniguchi *et al.*, 1965; Selby-Smith *et al.*, 1975). However, the mechanisms by which milk satisfies the requirements of lactic-acid bacteria for carbon dioxide, free amino acids and fermentable sugars are the most complex and widely-studied; they are therefore reviewed in detail in the following discussion.

2.1. Carbon Dioxide

Carbon dioxide has been shown to have both stimulatory and inhibitory effects on different lactic-acid bacteria. For example, it stimulates *Lb. bulgaricus* in continuous culture (Driessen *et al.*, 1982); both CO_2

Valerie M. E. Marshall and Barry A. Law

TABLE 2
Growth Requirements of Dairy Lactic-acid Bacteria

	Str. lactis	Str. cremoris	Str. diacetylactis	Str. thermophilus	Lactobacillus spp.	Leuc. cremoris
Amino acids						
Lysine	−	−	−	+	+	±
Leucine	+	+	+	+	+	±
Histidine	+	+	+	+	+	±
Valine	+	+	+	+	+	+
Cysteine	S	+	S	+	S	±
Aspartate	NI	NI	NI	+	+	+
Glutamate	+	+	+	±	+	±
Isoleucine	+	+	+	±	+	±
Tyrosine	NI	NI	NI	±	+	±
Methionine	+	+	+	±	+	±
Vitamins						
Vitamin B_{12}	+	+	+	+	−	NI
Biotin	+	+	+	+	+	NI
Niacin	+	+	+	+	+	NI
Pantothenate	+	+	+	+	+	+
Riboflavin	+	+	+	+	+	+
Thiamine	+	+	+	+	−	+
Pyridoxal	+	+	+	+	−	+
Folic acid	+	+	+	+	−	+
Organic acids						
Acetic acid	+	+	+	NI	NI	NI
Oleic acid	+	+	+	NI	S	NI
Orotic acid	NI	NI	NI	NI	S	NI
Formic acid	NI	NI	NI	NI	S	NI
Nucleic acids						
Hypoxanthine	S	−	−	NI	−	+
Adenine	S	S	−	NI	S	+
Guanine	S	−	−	NI	S	+
Thymine	S	−	−	NI	−	−
Thymidine	S	−	−	NI	−	−
Uracil	S	−	−	NI	S	+

+, essential for growth; ±, essential only in some strains; −, not required for growth; NI, not investigated; S, Stimulatory.

(~31 mg/kg) and formate (40 mg/kg) will produce a growth rate in milk similar to that obtained when grown in conjunction with *Str. thermophilus*. There is a recent report (Tinson *et al.*, 1982b) that *Str. thermophilus* produces more CO_2 than *Str. lactis* or *Str. cremoris* when grown in milk because this organism is capable of hydrolysing the urea present in milk to NH_3 and CO_2. Reiter & Oram (1962) found that *Str. lactis* ML3 had an absolute requirement for CO_2 in milk, whereas Oliver (1953) found that CO_2 interfered with the metabolism of *Str. lactis* subsp. *diacetylactis*. Later work however (Seitz *et al.*, 1963) showed that this organism was neither stimulated nor inhibited by exogenous CO_2 and the authors postulated that a high concentration of CO_2 forced the organism to become adaptively more homofermentative and to limit the heterofermentative CO_2-yielding reactions.

2.2. Carbohydrates

2.2.1. Mesophilic Cultures

Lactose. Str. cremoris and *Str. lactis* ferment lactose by the hexose diphosphate pathway to produce mainly lactic acid (Fig. 1). The key enzyme for this pathway is β-D-phosphogalactoside galactohydrolase (β-Pgal), first named by Laue & MacDonald (1968a,b), which cleaves lactose phosphate to produce glucose and galactose-6-phosphate. The reaction catalysed by this enzyme differs from that of β-D-galactoside galactohydrolase (β-gal) in that one of the monosaccharide products (galactose) is already phosphorylated. The galactose phosphate is metabolized to triose phosphate directly *via* the tagatose phosphate pathway, analogous to the fructose phosphate pathway from glucose. Following hydrolysis of lactose by β-gal, on the other hand, glucose is phosphorylated at the 6 carbon position and is metabolized *via* the Embden–Meyerhof pathway (Fig. 1). Many of the starter bacteria have been examined for β-gal and β-Pgal activity (Table 3). The predominant enzyme in the group N streptococci (except 7962) is β-Pgal. *Str. thermophilus* has only β-gal and does not convert the galactose moiety of lactose to lactic acid; instead this monosaccharide is released into the external medium (O'Leary & Woychick, 1976; Tinson *et al.*, 1982a). *Str. lactis* 7962 transports lactose into the cell as the free sugar and uses its β-gal to cleave it into two unphosphorylated monosaccharides (McKay *et al.*, 1969). Hexokinase phosphorylates the glucose at the carbon 6 position and it enters the Embden–Meyerhof pathway. Unlike *Str. thermophilus*, however, strain 7962 phosphorylates the

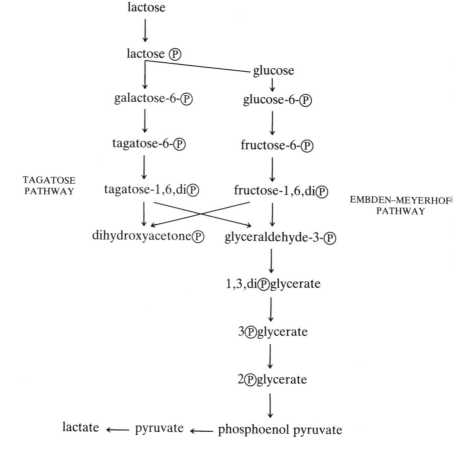

Fig. 1. The hexose diphosphate pathway.

galactose at the carbon 1 position because it has high levels of galactokinase enzyme. This sugar then enters the Leloir pathway and galactose-1-P is converted first to glucose-1-P and then to glucose-6-P. The enzymes of the Leloir pathway are present in the other group N streptococci but are possibly repressed when these organisms are grown on lactose (Lawrence *et al.*, 1976). The ability of the group N streptococci to ferment lactose by the hexose diphosphate pathway means that 90% of the lactose is fermented to lactic acid.

TABLE 3
Lactose-hydrolysing Enzymes in Starter Bacteria

Organism		β-gal	β-Pgal	Reference
Str. lactis	C10	14[a]	101[a]	
	7963	19	80	
	11454	1	118	
	7962	73	5	McKay *et al.* (1970)
Str. cremoris	9596	3	246	
	9625	1	151	
Str. lactis subsp.	DRC 3	3	254	
diacetylactis	11007	1	71	
Str. thermophilus	821	1·04[b]	0[b]	
	573	0·50	0	Tinson *et al.* (1982a)
	TS1	1·61	0	
Lb. bulgaricus		2363[c]	158[c]	
Lb. lactis		3948	102	Premi *et al.* (1972)
Lb. leichmanii		4539	235	
Lb. casei		0	412	

[a] Units expressed as nanomoles ONP formed min^{-1} mg cell dry wt^{-1}; [b] units expressed as μmoles ONP formed min^{-1} ml $culture^{-1}$; [c] units expressed as μmoles ONP formed min^{-1} mg enzyme $protein^{-1}$.

Lactose is phosphorylated by phosphoenol pyruvate (PEP) during translocation by the PEP-dependent phosphotransferase system (PEP:PTS; see Section 3) (McKay *et al.*, 1970). Phosphoenol pyruvate (PEP) and its enzyme pyruvate kinase occupy a pivotal role in the regulation of carbohydrate metabolism in these organisms (Thomas, 1976b; Thompson & Thomas, 1977; Thompson, 1978). Pyruvate kinase converts PEP to pyruvate which is subsequently converted to end metabolites and the phosphorylating ability of PEP is lost (although ATP is formed). All the phosphorylated intermediates of the hexose diphosphate pathway (Fig. 1) prior to 1,3-diphosphoglyceric acid are activators of pyruvate kinase (Thomas, 1976a). 2 Phosphoglyceric acid does not activate pyruvate kinase (Thompson & Thomas, 1977) and is the precursor of PEP. Control of pyruvate kinase therefore regulates the intracellular PEP 'pool' and hence couples sugar entry to sugar metabolism providing a tight link between substrate entry and its subsequent metabolism. Lactate dehydrogenase also controls carbohydrate metabolism. It is activated by fructose-1,6-diphosphate (Jonas *et al.*, 1972) and by tagatose-1,6-diphosphate (Thomas, 1975) ensuring conversion of pyruvate to lactate and limiting diversion of pyruvate into other products.

Galactose. Group N streptococci have the enzymic potential for galactose metabolism *via* two separate routes, i.e. tagatose-6-phosphate pathway and the Leloir pathway. The pathway which operates is dependent on the mechanism of uptake. If the PEP:PTS operates then the galactose appears within the cell as galactose-6-phosphate and is metabolized *via* the tagatose pathway. Before entry to the Leloir pathway free galactose must be transported *via* an ATP energized 'permease' and cytoplasmic phosphorylation occurs at the carbon 1 of the hexose.

Thomas *et al.* (1980) found that in *Str. cremoris* the tagatose pathway operated and agreed with Andrews & Lin (1976) who postulated that the scavenging power of the PTS was greater than the permease. However at low substrate concentrations, particularly in *Str. lactis* galactose metabolism may be *via* the Leloir pathway. Earlier these workers (Thompson *et al.*, 1978) had postulated that galactose utilization was under control by catabolite repression; glucose or subsequent metabolites may regulate the activities of the Leloir enzymes. In *Str. lactis* ML3, growth on galactose resulted in homolactic fermentation whilst growth of all other strains resulted in production of formate, acetate and ethanol as well as lactate (Thomas *et al.*, 1980). Reduced levels of both lactate dehydrogenase (LDH) activators and pyruvate-formate lyase activator appeared to be the main factors involved in diversion of pyruvate to other products. Le Blanc *et al.* (1979) also found that a *Lac* mutant of *Str. lactis*, deficient in galactose phosphotransferase system, produced more acetate and ethanol from galactose than did the parent strain. Growth of *Str. lactis* on low or limiting concentrations of carbohydrate (achieved under chemostat conditions) resulted in conversion of glucose into formate, acetate and ethanol. Similar studies with *Str. cremoris* showed this organism to be essentially homolactic (Thomas, 1979). A mutant of *Str. lactis* deficient in lactate dehydrogenase accumulates a large intracellular pyruvate pool. This results in conversion to diacetyl and acetoin (McKay & Baldwin, 1974). A similar situation was reported by Anders & Jago (1970) where diacetyl was produced by a strain of *Str. cremoris* whose lactate dehydrogenase was inhibited by fatty acids. An increase in pyruvate pool also occurs when *Str. lactis* subsp. *diacetylactis* metabolizes citrate with similar results.

Citrate. Under conditions in which organisms are grown in the presence of a 'normal' carbohydrate source (lactose, glucose, galactose), pyruvate must be converted to lactic acid to regenerate NAD from the NADH formed during breakdown of sugar to pyruvate. *Str. lactis* subsp. *diacetylactis* is unable to use citrate as a source of energy for growth, but will metabolize it and show an increase of 35% in specific growth rate (Harvey & Collins, 1963). The rate of entry of citrate is independent of growth rate and these authors postulated that citrate allowed the synthesis of a cell component that was produced only very slowly in its absence. The precursor of this cell component was probably lipid. Thus Harvey & Collins proposed the following:

$$citrate \rightarrow pyruvate \rightarrow lipid\ cell\ material$$
$$\downarrow$$
$$acetoin$$

However, more pyruvate was produced from the incoming citrate than was required for cell synthesis, so the excess was converted to acetoin.

A similar situation may exist for *Leuc. lactis* which will also utilize citrate in the presence of a metabolizable carbon source; Cogan *et al.* (1981) observed only trace amounts of diacetyl and no acetoin production when this organism was given citrate. It was only after citrate was depleted that the maximum amount of acetoin was produced. The pathway of citrate utilization is shown in Fig. 2. The four enzymes, citrate lyase, acetolactate synthase, diacetyl reductase and acetoin reductase have been characterized in *Str. lactis* subsp. *diacetylactis*, *Leuc. lactis* and *Lb. plantarum* (Cogan, 1981; Mellerick & Cogan, 1981).

Citrate lyase requires Mg^{2+} or Mn^{2+} (Harvey & Collins, 1961), has an optimum pH of 7·0 and is constitutive in *Str. lactis* subsp. *diacetylactis* but inducible in *Leuc. lactis*. In the former organism however, activity is increased as growth rate is increased (whether citrate is present or not). Acetolactate decarboxylase also requires Mg^{2+} or Mn^{2+}, has an optimum pH of 6·5 in acetate buffer, but exhibits reduced activity in phosphate buffer. Activity in *Str. lactis* subsp. *diacetylactis* decreased as growth increased, with slightly less decrease in the presence of citrate. It may therefore be partly inducible. This enzyme is also inducible in *Leuc. lactis*. Diacetyl synthase and acetoin synthase both have optimal activity at pH 5·5. Activities increased as growth rate increased in the absence of citrate. However, with *Str. lactis* subsp. *diacetylactis* the presence of citrate repressed their synthesis whereas in *Leuc. lactis* these enzymes were constitutive.

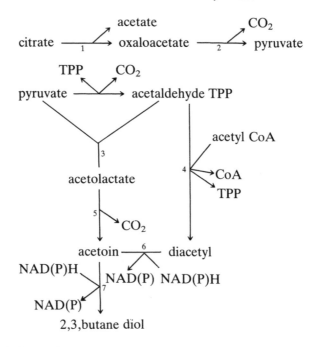

Fig. 2. Pathway of citrate metabolism: 1, citrate lyase; 2, oxaloacetate decarboxylase; 3, acetolactate synthase; 4, diacetyl synthase; 5, acetolactate decarboxylase; 6, diacetyl reductase; 7, acetoin reductase.

The ability to ferment citrate has long been known to be unstable. In 1980 Kempler & McKay demonstrated that citrate-negative mutants of *Str. lactis* subsp. *diacetylactis* lacked a 5·5 megadalton plasmid and postulated that citrate utilization was plasmid linked. It may be however that it is only the citrate permease enzyme which is plasmid carried (see Kempler & McKay, 1981).

2.2.2. Thermophilic Cultures
Detailed studies on carbohydrate metabolism have not been made on *Str. thermophilus*. It lacks the β-Pgal enzyme (although one strain does have some activity (Hemme *et al.*, 1981a)) and is therefore expected not to have the enzymes of the tagatose pathway. β-Gal is synthesized constitutively except for strain ST4 where it is inducible (Reddy *et al.*, 1973). It is not repressed when the organism is grown on glucose as the sole carbon source and is increased 2–4 fold on introduction of lactose

or galactose (Tinson *et al.*, 1982a). Disaccharides are metabolized in preference to monosaccharides (O'Leary & Woychick, 1976) and when grown on lactose, galactose accumulates in the medium. The organism only metabolizes galactose *via* the Leloir pathway when lactose is limiting. Similar sequential sugar metabolism has also been found for *Str. lactis* (Thompson *et al.*, 1978). Unlike this latter organism however, *Str. thermophilus* may excrete glucose as well as galactose during growth on lactose (Thomas & Crow, 1983). However, this may be a consequence of release of β-galactosidase from the cells during growth.

As lactate is the primary end metabolite it would be expected that lactate dehydrogenase would be under a similar control to that of *Str. lactis*, but Hemme *et al.* (1981a) have shown that this enzyme is inhibited by fructose diphosphate at concentrations greater than 1 mM. It may be significant that pyruvate is also converted to acetaldehyde by the pyruvate dehydrogenase. This organism lacks alcohol dehydrogenase and as a consequence acetaldehyde is excreted into the external medium where concentrations can become as high as 30 ppm (Marshall *et al.*, 1982). The central role of acetaldehyde in metabolism is reviewed by Lees & Jago (1978a,b).

Metabolism of the lactobacilli is also homolactic and β-gal is the predominant enzyme (Premi *et al.*, 1972) with lactate the final product. Many lactobacilli are capable of producing acetaldehyde and alcohol as end metabolites (Keenan & Lindsay, 1966). Like *Str. thermophilus*, *Lb. bulgaricus* lacks alcohol dehydrogenase and will not convert acetaldehyde to ethanol. The acetaldehyde, however, may be the result of threonine metabolism in this organism (Lees & Jago, 1976).

2.3. Proteins

2.3.1. Amino Acids and Peptides in Nutrition

All dairy lactic-acid bacteria either require, or are stimulated by, amino acids, yet free amino acid concentrations in milk are not sufficiently high to permit commercially useful growth and acid production (Table 4). Detailed experimental data about these requirements are only available for some of the group, the most definitive studies having been carried out with group N streptococci (Reiter & Oram, 1962; Law *et al.*, 1976) and *Str. thermophilus* (Shankar, 1977; Bracquart & Lorient, 1979). Qualitative data on lactobacilli (Morishita *et al.*, 1981) indicate that this group has the most extensive requirements. The dairy lactic acid bacteria overcome the nutritional problem pre-

TABLE 4

Minimum Concentrations of Amino Acids required by Group N Streptococci and *Streptococcus thermophilus* for Maximum Growth in Defined Media + Lactose After 17 h at 24°C (data from Law *et al.*, 1976; Shankar, 1977 and Mills & Thomas, 1981)

	Concentration required in medium for maximum growth ($\mu g\ ml^{-1}$)				Concentration of free amino acids in milk ($\mu g\ ml^{-1}$)
	Str. lactis	Str. lactis *subsp.* diacetylactis	Str. cremoris	Str. thermophilus	
Glu	77	87	70	150	35·9
Leu	41	37	32	n.e.	1·2
Ile	33	30	32	n.e.	0·8
Val	27	30	41	n.e.	2·6
Arg	37	36	39	n.e.	1·6
Cys	s	s	27	80	n.d.
Pro	n.e.	n.e.	38	n.e.	8·8
His	23	24	14	60	2·8
Phe	31	n.e.	6	n.e.	n.e.
Met	22	21	11	n.e.	n.d.

n.d., not detected; n.e., not estimated; s, stimulatory.

sented by the low concentrations of free amino acids in milk using a complex combination of proteinases, peptidases and transport systems to make protein-bound amino acids available for growth.

Like other bacteria, the lactic acid bacteria can actively transport amino acids and peptides across the cell membrane into the cell against a concentration gradient (Law & Kolstad, 1983). The free amino acids normally present in fresh milk are sufficient to support the growth of *Str. cremoris* to cell densities corresponding to 8–16% of those found in coagulated (fully-grown) milk cultures (Mills & Thomas, 1981). Peptides of molecular weight (mol. wt.) <1500 provide a further source of amino acids, though milk proteins become important at high cell densities. Amino acid and peptide uptake in lactobacilli, leuconostocs and group N streptococci are mediated by separate systems, and the streptococci also have distinct dipeptide and oligopeptide transport (Law, 1978). The size exclusion limit for uptake by the latter system corresponds to peptides containing between three and seven residues (estimated average size in mixed peptides; Law *et al.*, 1976) or five residues (estimated with a single peptide series; Rice *et al.*, 1978). There is strong presumptive evidence that *Str. lactis* can utilize peptides *via* cell wall-bound peptidases as well as transporting them intact.

For example, although uptake competition between peptides can be demonstrated in *Str. lactis* it is invariably weaker than that measured in *Str. cremoris* (Law, 1978) and is not seen at all in longer-term growth response experiments (Law, 1977). These observations are consistent with the findings that whole cells of *Str. lactis* hydrolyse peptides under non-transporting conditions and that peptidases are released when the cells are protoplasted or osmotically shocked under a variety of conditions (Law, 1979; Kolstad & Law, unpublished).

The significance of the uptake and utilization of amino acids and peptides in *Str. thermophilus* is probably more complex than is the case with group N streptococci. The thermophilic streptococci not only have an absolute requirement for some amino acids but are also stimulated by non-essential amino acids (valine, methionine, leucine, tryptophan) to produce more acid than normal in milk (Shankar, 1977; Bracquart & Lorient, 1979). Presumably, their rates of cellular synthesis are rate-limiting for growth in the presence of 'normal' amounts in milk. If these amino acids are supplied as peptides, similar stimulation can be observed (Desmazeaud & Hermier, 1972; Shankar, 1977). Stimulatory peptides isolated from milk cultures or casein digests by these authors would be too large to be transported by any of the known uptake systems in lactic acid bacteria but *Str. thermophilus* probably has surface-bound peptidases which reduce them to a more manageable size (Shankar, 1977).

2.3.2. Proteolytic Systems

The detailed studies of the last decade have given an insight into the complexity of the proteolytic 'equipment' of the lactic-acid bacteria. This complexity is seen not only in the numbers and types of different proteinases and peptidases but also their sub-cellular distribution (Fig. 3).

The first enzymes to make contact with milk proteins are the extracellular proteinases. Many of the dairy strains produce such enzymes but controversy surrounds some of the literature reports as to the true nature of the activities detected. For example, Williamson *et al.* (1964) claimed that *Str. lactis* secreted a cell-free extracellular proteinase, but the criteria for conferring true extracellular status (Pollock, 1962) on the enzyme were not applied, especially in relation to the lack of any assessment of cell lysis or leakage. Also, a later report by Cowman & Speck (1967) showed that the same organism produced an intracellular proteinase with some properties in common

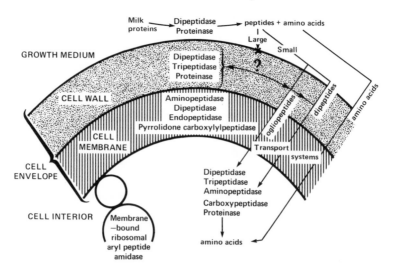

Fig. 3. Schematic representation of the interrelationships between proteolytic systems and transport systems in lactic-acid bacteria. Arrows indicate the direction of utilization from protein to amino acids. The double headed arrow indicates that there may be functional interactions between cell wall enzymes and transport systems. See text for details.

with the supposed extracellular enzyme. A further source of cell-free extra-cellular proteolytic activity (especially in cultures grown in non-milk media low in Ca^{2+}; Mills & Thomas, 1978) could be the loosely-bound cell wall-located proteinase(s) first described by Thomas et al. (1974). Only one report (Exterkate, 1979a) suggests that group N streptococci may secrete a distinct cell-free extracellular proteinase and this is based on the different Ca^{2+} requirements of the cell-bound and cell-free proteolytic activities.

The cell wall location of a high proportion of the group N streptococcal proteinase activity is well established in that it can be detected in spheroplast supernatants containing lysozyme- or lysin-released cell wall components in the absence of intracellular marker enzymes (Thomas et al., 1974; Exterkate, 1975). The amount of activity recovered in this way is approximately equal to that measurable with whole cell suspensions incubated with casein. No purification and characterization of a cell wall proteinase have been reported, but inhibitor studies on crude preparations suggest that the Str. lactis enzyme is a metalloproteinase. Exterkate (1976) distinguished three types of proteinase activity according to their temperature and pH optima and showed that there are wide strain variations in the numbers and combinations of these enzymes. Multiple proteinases in *Str. lactis* have recently been demonstrated by zymograms of lysozyme extracts from milk-grown cells. At least four bands of caseinolytic activity were separated and shown to be distinct from the intracellular activity which is apparently due to only one proteinase (Cliffe & Law, unpublished data).

Although relatively little is known about the proteinases themselves, the factors controlling their synthesis, activity and attachment to the cell wall have been described. For example, *Str. cremoris* and *Str. lactis* release part of their proteolytic activity without lysing if they are suspended in buffer (Mills & Thomas, 1978). This phenomenon is temperature- and pH-dependent, and suppressed by added Ca^{2+}, suggesting an involvement of this ion in enzyme attachment. However, these results were obtained under experimental conditions which may not be relevant for understanding the normal function of Ca^{2+} in cells when they are actively producing proteinases under the influence of constituents in their growth medium. Perhaps this is why the later study of Exterkate (1979a) produced a different conclusion concerning the role of Ca^{2+}; a dependence on Ca^{2+} for accumulation of active proteinase in cell walls was observed in growing cultures but the ion stabilized the enzyme molecules in an active configuration, rather than binding them to the cell wall. In contrast to Mills & Thomas (1978), Exterkate (1979a) observed that actively growing or metabolizing cells released a constant amount of proteinase into the medium, irrespective of its Ca^{2+} content; only the enzyme in the cell was affected by the Ca^{2+}.

The *de novo* synthesis of cell wall proteinase is subject to regulation by amino acids or peptides at the level of mRNA translation (Exterkate, 1979a). Proteinase-specific mRNA appears to be either inherently long-lived or stabilized by the proximity of the cell membrane to the ribosomes (cf. Shires *et al.*, 1974). Low concentrations of casein-derived amino acids do not prevent proteinase synthesis but cells respond to high extracellular concentrations by reducing the rate of enzyme synthesis until cellular metabolism has used up this external N source; proteinase synthesis then resumes in order to ensure a further supply of amino acids from proteins in the medium. Both the effect of Ca^{2+} on proteinase stability and the amino acid/peptide effect on its synthesis may explain the well-documented failure of cells which have been grown in non-milk media to grow rapidly in milk on sub-culture.

Zymograms of cell wall proteinases from *Str. lactis* cultures harvested from a complex, nutritionally rich medium (M17; Terzaghi & Sandine, 1975) illustrate the cause of this phenomenon when it is compared with one from milk-grown cells. Both media yield proteolytically active cells but the M17-grown cells are lacking two of the proteinase bands found in milk-grown cells (Cliffe & Law, unpublished data). This suggests differential repression or destabilization and warrants further study.

Populations of group N streptococci undergo spontaneous proteinase loss at high frequency, suggesting that the enzymes are encoded on extra-chromosomal DNA. Plasmid linkage has long been suspected from evidence gained in curing experiments (see review by Davies & Gasson, 1981). Unequivocal, direct evidence to show that the proteinase gene(s) is carried on a plasmid has only recently been published after many confusing and conflicting reports assigning it to plasmids ranging from 10–30 megadaltons (Mdal). Gasson (1983) used protoplast regeneration to produce variants of *Str. lactis* which only carried a 33 Mdal plasmid, yet retained proteinase activity. Loss of this plasmid results in the loss of several cell wall-bound proteinases detectable in the zymograms of cell wall extracts (Cliffe & Law, unpublished data).

Of the other lactic-acid bacteria, lactobacilli and leuconostocs probably have cell wall-bound proteinases as evidenced by the hydrolysis of milk proteins by whole cell suspensions (Searles *et al.*, 1970). Argyle *et al.* (1976) and Chandan *et al.* (1982) showed that the activity of *Lb. bulgaricus* was released from the cells by treatment with lytic enzymes or by osmotic shock. However, the precise location of the proteinase cannot be deduced from these studies; no estimate of intracellular enzyme leakage was reported, and its recorded properties may have been those of a mixture of proteinases from various other cellular locations. Despite these reservations there is little doubt that *Lb. bulgaricus* produces an extracellular proteinase and it is understood that the peptides which it releases from milk proteins act as stimuli to the growth of *Str. thermophilus* in yogurt cultures (Shankar & Davies, 1978; Hemme *et al.*, 1981b).

A cell wall-bound proteinase has been demonstrated in *Lb. helveticus* and its location is supported by evidence that it is released from whole cells without any leakage of intracellular enzymes (Vescova & Bottazzi, 1979). However, the enzyme was not purified or characterized.

There is an increasing body of evidence suggesting that in some lactic-acid bacteria, protein and peptide utilization is aided by extracellular peptidases as well as proteinases. Law (1977, 1979) and Shankar & Davies (1978) recovered dipeptidase activity from culture supernatants of *Str. lactis, Str. cremoris* and *Str. thermophilus* after they had reached the stationary growth phase. On the available evidence it is not possible to decide whether these represented truly extracellular enzymes, though the observations of Law (1979) that only one type of activity is found in the medium, whereas many peptidases are confined to the intracellular location, suggest selective release of the dipeptidase. Against this, it is clear that the 'extracellular' dipeptidase shares several properties with the intracellular true dipeptidase in *Str. lactis* and *Str. cremoris.*

Evidence for the existence of a distinct cell wall-bound peptidase in *Str. lactis* (but not *Str. cremoris*) is more convincing; Sørhaug & Solberg (1973) treated acetone-dried whole cells with trypsin and released a peptidase whose substrate profile suggested that it was a different enzyme from those released by mechanical disruption. It was not decided whether this peptidase was cell wall or membrane-bound and no data on intracellular enzyme leakage were reported. Law (1977, 1978) noted that although *Str. lactis* utilized peptides containing essential amino acids, its growth was not inhibited by other peptides containing non-essential amino acids competing for the same transport system. Also, uptake competition between ^{14}C-labelled and unlabelled peptides was weak in *Str. lactis* compared with *Str. cremoris,* suggesting that a proportion of the peptide was hydrolysed outside the cell. Taken together with the observation that starved, non-transporting whole cells of *Str. lactis* hydrolysed peptides, this presumptive evidence supports the idea of a cell wall peptidase. In addition, a low mol. wt. (26 000) EDTA-sensitive di-/tripeptidase is released from *Str. lactis* after treatment with lysozyme or suspension in Tris-buffer, under conditions in which the release of intracellular marker enzymes is only 1–2% (Law, 1979; Kolstad & Law, unpublished data).

The report of surface-bound peptidases of *Str. thermophilus* (Shankar & Davies, 1978) was not supported by tests for the possibility of cell leakage or lysis, but the existence of such enzymes would help to explain the stimulation of this organism by *Lb. bulgaricus* in yogurt cultures. El Soda *et al.* (1978b) did not find peptidases in the solubilized cell wall fraction of *Lb. casei*, but Eggiman & Bachmann (1980) have purified a surface-bound aminopeptidase from *Lb. lactis*. This enzyme has a broad specificity, hydrolysing di- and tripeptides as well as aminopeptidase substrates. It shares this and many other properties with intracellular aminopeptidases of lactic acid bacteria and therefore convincing evidence is needed to distinguish it unequivocally as a truly extracellular enzyme.

Further involvement of peptidases in peptide uptake by group N streptococci is suggested by several reports of membrane-bound peptidases. For example, the endo- and aminopeptidase activities of osmotically lysed *Str. cremoris* HP are found in the cell fraction sedimenting at approximately 48 000g and may be involved in peptide utilization by this organism if this fraction is assumed to contain cell membranes (Exterkate, 1975). Exterkate (1977, 1979b) later showed that the kinetic properties and the interactions of the aminopeptidases with solvents and detergents were consistent with a membrane location. The physiological function of these 'membrane-bound' peptidases remains unclear, partly because they were not tested against unsubstituted peptide substrates which the organism could be expected to encounter. Exterkate (1982) suggested a functional relationship between an endopeptidase and aminopeptidase protein whereby the activation of one is effected by the other. It would be desirable to know the natural substrates of these enzymes in order to determine their function. Indeed, it is possible that the 'endopeptidases' (P_{37} and P_{50}) described by Exterkate (1975) are aryl-peptide amidases (EC 3.5.1) and have no general peptidolytic or proteolytic activity.

Such enzymes are present in the ribosomal fraction of *Lb. casei* (El Soda & Desmazeaud, 1981) and may function to cleave specific bonds in the signal sequence of newly synthesized proteins. These sequences are a vital part of the extrusion mechanism for cell envelope proteins emerging from membrane-bound ribosomes (Garnier *et al.*, 1980), and the nature of the fraction bearing the P_{37} and P_{50} activities is not inconsistent with such a location and function for these enzymes. Although Exterkate (1975) had assumed that the amino- and endopeptidases were located on the outer surface of the membrane, this conclusion was only based on the ability of whole cells to degrade the appropriate substrates. While this may be valid, the possibility that the derivatized peptides could cross the cell envelope passively was not investigated. Law (1979) used unsubstituted peptides to show that di- and tripeptidases were present in both the 'particulate' (100 000g pellet) and 'cytoplasmic' fractions of osmotically lysed cells but, contrary to Exterkate's findings, the leucine aminopeptidase was largely cytoplasmic in *Str. cremoris* NCDO 1196 and 2016, and completely so in *Str. lactis* NCDO 2017. In addition, the use of dipeptide substrates revealed that a peptidase(s) was released from *Str. cremoris* NCDO 1196 by lysozyme or by osmotic shock, in the absence of intracellular marker enzyme leakage (Law, 1979; Kolstad & Law, unpublished data). The *in vivo* location of this enzyme has not been established but studies with whole cells suggest that it does not have access to peptides in the growth medium, unlike the surface-bound peptidase of *Str. lactis*.

It is unlikely that the strains of *Str. cremoris* investigated by Law (1977, 1978) for peptide utilization had peptidases on the outside of their membranes since competition experiments, both in terms of growth response and short-term uptake kinetics, proved that peptides were transported intact into cells by energy-dependent systems. If any hydrolysis occurred, it must have done so within the membrane, on its inside surface, or completely intracellularly. Intramembrane hydrolysis of peptides could conceivably play a role in peptide uptake (Payne, 1980; Hermsdorf & Simmonds, 1980) but a membrane-located enzyme active against unsubstituted peptides would be required to fulfil such a function. Kolstad & Law (unpublished data) isolated a Triton-activated dipeptidase of very limited specificity from purified membranes of *Str. lactis* and considered it to be distinct from the 'cell wall' and cytoplasmic peptidases on the basis of localization, electrophoretic mobility and specificity against a range of peptides. Comparative peptide specificity is summarized in Table 5. However, although this enzyme does hydrolyse unsubstituted peptides, its very narrow specificity (chiefly for Ala–Phe) would restrict its usefulness as a general mediator in peptide uptake.

Once inside the cell, intact peptides can be hydrolysed by the wide range of cytoplasmic peptidases. The group N streptococci contain peptidases whose spectrum of bond specificities is probably wide enough to ensure the complete release of all amino acids from casein-derived peptides (Mou *et al.*, 1975). Although these authors did not characterize any of the enzymes, they concluded that crude cell-free extracts of *Str. lactis* subsp. *diacetylactis* and *Str. cremoris* contained five different peptidases based on the substrates which were hydrolysed; they found di- and tri-peptidases, aminopeptidase-P, proline iminopeptidase and general aminopeptidase activities. Electrophoretic zymogram studies of crude sonic extracts of the three species of group N streptococci revealed between four and nine different peptidases (Sørhaug and Solberg, 1973; Sørhaug & Kolstad, 1981). Law (1979) found only three readily distinguishable peptidases in *Str. lactis* and *Str. cremoris* after the cells had been osmotically lysed and the extract

TABLE 5
Comparative Peptide Specificities of Cell Wall, Cell Membrane, and
Cytoplasmic Fractions of *Streptococcus lactis*

	Cell wall	Membrane (in Triton)	Cytoplasmic
Leu–Gly	++++	+	++++
Pro–Phe	+++	–	++
Phe–Pro	+	–	+
Ala–Phe	++	++	+
Ala–Trp	+	–	+
Leu–Ala	++++	–	+++
Ala–Leu	++++	+	+++
Pro–Leu	+	–	–
Pro–Met	++++	+	++
Met–Pro	–	–	–
His–Phe	++	–	+
Gly–Phe	+++	+	++
Pro–Ile	+	–	++++
Val–Pro	–	–	+
Trp–Ala	+	–	–
Leu–Leu–Leu	–	–	+
Ala–Leu–Gly	–	–	–

–, no bands; +, ++, +++, ++++, band intensity from weak to
strong. Peptidases were separated by polyacrylamide gel electrophoresis
then detected as coloured bands by a modification of the method of
Lewis & Harris (1967).

centrifuged at a sufficiently high speed to ensure that all particulate
fractions had been sedimented. This suggests that *in vivo* there are
fewer cytoplasmic peptidases than previous studies of sonicates or
homogenates would indicate. However, this supposition requires con-
firmation with a wider range of peptides and peptide derivatives as
substrates.

Like the group N streptococci, *Str. thermophilus* has intracellular
peptidases capable of hydrolysing a wide range of substrates. Desmazeaud & Juge (1976) used electrophoretic zymograms to demonstrate the presence of two aminopeptidases and three dipeptidases,
though some of the latter activity was attributable to one of the
aminopeptidases. Peptidases in lactobacilli have been studied in detail
recently and they appear to have a wider range of activities than the
lactic streptococci. *Lb. casei*, for example, has a broad-specificity true
dipeptidase, an aminopeptidase, a narrow-specificity carboxypeptidase

and an apparent endopeptidase, later identified as an aryl peptide amidase (El Soda *et al.*, 1978a; El Soda & Desmazeaud, 1981). A carboxypeptidase was also found in one strain of *Lb. helveticus* but not in *Lb. lactis*, *Lb. bulgaricus* and lactobacilli of the betabacterium group (El Soda & Desmazeaud, 1982; El Soda *et al.*, 1982). Amidase activity similar to that in *Lb. casei* was subsequently detected in *Lb. brevis* and *Lb. fermentum*, but not *Lb. helveticus*, *Lb. bulgaricus*, *Lb. lactis*, *Lb. acidophilus*, *Lb. buchneri* or *Lb. cellobiosus*.

3. MECHANISMS OF NUTRIENT UPTAKE

In energy-independent solute transfer, a substrate is transported into the cell to a point where the concentration on each side of the membrane is the same. Thereafter exchange can occur, but there will be no net flux in either direction. The substrate can only diffuse down its own concentration gradient. Energy coupling is required to allow movement against the concentration gradient (i.e. active transport) and so produce intracellular accumulation. In this situation, it is the position of the final equilibrium which is being altered and, as in enzyme-catalysed reactions, the transport is coupled to another reaction, in itself characterized by a decrease in free energy.

Depending upon the solute and its transport system, the active transport of sugars and amino acids across the bacterial cell membrane occurs at the expense of three energy sources which are not mutually exclusive: (a) oxidative phosphorylation and respiration-linked electron flow, (b) membrane bound ATPase, and (c) the phosphoenolpyruvate:phosphotransferase system (PEP:PTS). The energy produced by (a) and (b) contributes to the maintenance of an energized membrane state or membrane potential. In bacteria which lack the respiratory chain (this includes the lactic acid bacteria) (a) cannot operate as a mechanism for generating a membrane potential. In these bacteria a membrane-bound ATPase (adenosine-5'-phosphatase, EC 3.6.1.3) therefore has a very important role.

ATPase is proton translocating; hydrolysis of ATP is coupled to efflux of protons, generating osmotic energy *via* a membrane potential and pH gradient (together making up a proton-motive force; Mitchell, 1963; 1970) where the interior of the cell is negative and alkaline. The reversibility of the ATPase means that the converse is also true; ATP synthesis can be driven by the proton motive force. Indeed, this has been demonstrated in *Str. lactis* (Maloney & Wilson, 1975; Maloney, 1977); when a proton-motive force was artificially imposed on the membrane protons moved inward and ATP was synthesized. However, Otto *et al.* (1983) have used membrane vesicles to investigate the requirement for ATP in the generation of an electrochemical proton gradient in *Str. cremoris* and show that under these conditions ATP hydrolysis is not an obligatory requirement. Otto *et al.* (1982) suggest that lactate efflux contributes significantly to energy requirements in group N streptococci. Nevertheless, there is a universal requirement for an osmotic energy source stored as a trans-membrane potential and pH gradient (chemiosmotic coupling), for subsequent use in energy re-quiring reactions.

The 'permease' model for solute translocation may be interpreted in terms of chemiosmotic energy coupling. *Str. faecalis* accumulates ami-no acids in response to the proton motive force (Asghar *et al.*, 1973; Thompson, 1976) in much the same way as *Staph. aureus* (Niven & Hamilton, 1973; Niven *et al.*, 1973). Acidic amino acids enter in response to a pH gradient because the interior of the cell is alkaline. Alkaline amino acids are positively charged and enter in response to a membrane potential difference where the interior of the cell is nega-tive. Neutral amino acids are co-transported with protons in a similar way to the 'permease' model for galactoside and lactose uptake in response to the total proton-motive force. The solutes are all translo-cated down an ion gradient. Further evidence for interpreting the permease in terms of chemiosmotic coupling came from studies of lactose uptake by *Escherichia coli* which is accompanied by an influx of protons, a flux which is in turn accompanied by an equal and opposite efflux of potassium ions (West & Mitchell, 1973). In this paper the flux was equated with the M protein of the permease described by Fox & Kennedy (1965). Primary solute transport can be discussed in general terms where a solute is driven by an energy source. Secondary solute transport must confer a specificity and requires the existence of trans-port proteins.

The PEP:PTS catalyses transport and phosphorylation of hexoses at the expense of PEP, requiring 3 or 4 proteins for the sequence. These proteins have been well characterized and their properties and function are reviewed by Robillard (1982). The initial step is the EI-catalysed conversion of PEP to pyruvate with phosphorylation of a second protein, HPr. EI is a metal-requiring hydrophobic enzyme which is active as a dimer but only carries one phosphoryl group. Transport and phosphorylation of the sugar are catalysed by a membrane bound protein (EII) *via* a phosphoenzyme intermediate (EIII) which may or may not be membrane bound but which is sugar specific and has a regulatory role.

In *E. coli* the PEP hexose transport system is sensitive to the presence of an electrochemical proton gradient (Reider *et al.*, 1979). The proton motive force inhibited transport. Otto *et al.* (1983) studied the growth of *Str. cremoris* in a chemostat where growth rates could be regulated and reported an unexpected observation: that PMF increased at low growth rates. Should the PMF be capable of inhibiting growth rates in *E. coli* and in *Str. cremoris*, then the results obtained in these chemostat studied may be a consequence of limited carbon supply.

Any model proposed for an uptake mechanism must include a directional component. In the above model, amino acids and sugars move inwards in response to a gradient, but how does the PEP:PTS translocate solutes? Hamilton (1975) in his excellent review has compared the cytoplasmic phosphorylation of glucose with the phosphorylation of the same sugar during transport by the PEP:PTS. The reactions of the cytoplasmic enzymes, pyruvate kinase and hexokinase are coupled to high energy phosphate bond hydrolysis in the same way as the reactions of enzymes I and II of the PEP:PTS (Fig. 4). In Fig. 4(a) glucose may approach the enzyme surface from one direction and leave as glucose-6-phosphate in another direction, but location is unimportant in this scheme as all components are cytoplasmic. This is not the case in Fig. 4(b). Enzyme II is membrane bound, glucose is extracellular and approaches from the outside medium. After phosphorylation it leaves the enzyme and becomes intracellular: a directional component has been imposed by the presence of the membrane. A pre-requisite of this model is that in normal circumstances the PTS scheme (b) cannot behave like scheme (a), i.e. it cannot convert intracellular glucose to glucose-6-phosphate.

Changes in K_m values of transport proteins in response to the proton motive force have been observed both in prokaryotes (Kaczorowski *et al.*, 1979) and in eukaryotes (Vignais, 1976; deMais & Vianna, 1979). In each case changes in affinities of one or two orders of magnitude have been reported. It is assumed that these affinity changes have a functional significance. Konings & Robillard (1982) suggested that enzyme II can exist in two forms: a high affinity state and a low affinity state, and the physical mechanism controlling the change of affinity is a redox process involving a dithiol–disulphide interchange. The reduced

a)

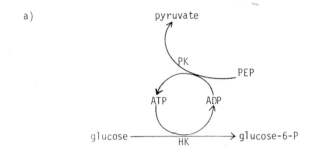

b)

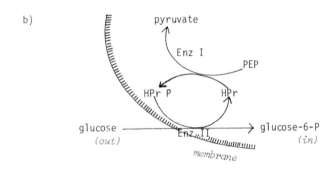

Fig. 4. Transfer of high-energy phosphate bond from phospho-enol-pyruvate (PEP) (a) by cytoplasmic enzymes; (b) by membrane bound enzymes. The reaction has now become directional. PK, pyruvate kinase; HK, hexokinase; HPr, heat stable protein; Enz I, Enzyme I; Enz II, Enzyme II complex.

state of the carrier is the low K_m (high affinity) form and the oxidized state is the high K_m (low affinity) form. The changes observed on altering the redox state are the same as the changes in K_m on establishing a proton motive force across the membrane (Robillard & Konings, 1981; Konings & Robillard, 1982).

There is evidence that both mechanisms are also operating in *Str. lactis*. The presence of β-Pgal suggests that lactose is transported by the PEP:PTS system and first appears in the cell as lactose phosphate (McKay *et al.*, 1969). Galactose, however, is transported by the PEP:PTS and by energy provided by the proton motive force (Kashket & Wilson, 1972; Thompson, 1980; Park & McKay, 1982). Thompson (1980) also compares the affinity of the two mechanisms for galactose with the chemiosmotic coupling being ten-fold greater than the PEP:PTS.

The work of Thompson & Thomas (1977) leads to some speculation about transport as a site of cellular control in *Str. lactis*. In the absence of an energy source galactose-grown cells of strain ML3 rapidly accumulated a galactoside. The uptake did not appear to be a consequence of the proton motive force, as dissipation of the proton gradient did not prevent accumulation. Analysis revealed high intracellular levels of PEP and 2-phosphoglycerate and of 3-phosphoglycerate (Thompson, 1978) but absence of phosphorylated metabolites of carbohydrate catabolism. These latter metabolites are activators of pyruvate kinase (Thomas, 1976b) so their absence ensures the presence of PEP. Mason *et al.* (1981) also found a high intracellular pool of PEP in starved cells of *Str. lactis*. The PEP is thus available for active transport as soon as a sugar is provided for the starved cells.

The involvement of cyclic AMP (cAMP) in control mechanisms should also be considered. Elevated cAMP has recently been shown to enhance lactic acid production by *Str. lactis* (Racliff & Talbart, 1981). An increase in intracellular cAMP was also shown by these authors to inhibit glucose repression (Racliff *et al.*, 1980). The role of cAMP as a control device in the synthesis of some transport enzymes in *E. coli* has been discussed by Peterkovsky (1981), but Ullmann & Danchin (1982) question this role and suggest that catabolite repression may not be a consequence of low cAMP, rather that low cAMP is a result of catabolite repression. This conclusion is echoed by Chassy & Thompson (1983) for *Lb. casei*; control of sugar transport in this organism cannot be mediated by this nucleotide since members of this genus do not produce cAMP (Sahyoun & Durr, 1972).

4. GROWTH IN MILK

Starter cultures must be able to multiply from 10^6–10^7 colony-forming units (cfu) ml^{-1} milk to 10^8–10^9 cfu g^{-1} curd (or other product) in a few hours in order to produce enough lactic acid and aroma compounds for the complete conversion of milk to fermented dairy products. Their ability to do this depends not only on nutritional factors but also on other environmental parameters such as temperature, culture pH and the presence of other organisms in the milk. These factors are now considered.

4.1. Temperature of Fermentation

The optimum temperature for growth of mesophilic starters lie between 25 and 30°C. At 30°C the group N streptococci have a mean generation time in milk of 60–70 min and grow to a maximum cell density of 0·5 mg dry weight of bacteria ml^{-1}. In complex broth media the same organisms have doubling times of 35–40 min (Lawrence & Thomas, 1979). Growth is limited as a result of low pH (4·5). Commercial milk cultures of the cheese starter strains are generally prepared at slightly sub-optimal temperatures (22–25 °C) in order to balance high cell numbers against over-acidification. Single strain starter cultures grown in separated milk at 25°C reach maximum cell numbers at pH 4·65 and this value (corresponding to about 0·7% lactic acid) is usually taken as the optimum for good subsequent acid development in cheese vats (Pearce *et al.*, 1973). If cultures are allowed to develop more acidity, their lag phase, on subculture to cheesemilk, is lengthened, presumably due to damage to cells by low pH and lactic acid. An evaluation of this effect on commercial mixed strain mesophilic starters was reported by Ross (1980).

During some cheesemaking processes (e.g. Cheddar) the group N starter streptococci are required to produce lactic acid at temperatures of up to 40°C. At such superoptimal temperatures, growth and acid production become uncoupled, especially in *Str. cremoris*, which is more temperature sensitive than *Str. lactis* (Breheny *et al.*, 1975). The kinetic response of *Str. cremoris* HP cultures to growth at superoptimal temperatures was reported by Franks *et al.* (1980). They demonstrated that the catabolic and anabolic activities of the cell decay at different rates as temperature increases, so that populations develop which contain increasing numbers of organisms capable of acid production, but not growth and multiplication. When a mathematical model based on this mechanism was tested experimentally, a delay in death was apparent in response to transient temperature rises. Using this model to predict acid production and population growth in cheese vats, Franks *et al.* (1980) were able to suggest a stepped-delay temperature programme which would allow the use of smaller starter culture inoculi to achieve normal rates and levels of acidity development. The concept of using minimal numbers of starter cells to acidify Cheddar cheese curd also has important application in the avoidance of the bitter defect in mature cheese (see Chapter 7).

The thermophilic starters, *Str. thermophilus* and the lactobacilli, are more acid tolerant and will reduce the pH to 4·1 in the case of *Str.*

thermophilus and to 3·8 in the case of *Lb. bulgaricus* when grown in milk at 42°C (Marshall *et al.*, 1982). Molar growth yields for a number of lactobacilli may be temperature-dependent. For example, those of *Lb. lactis* and *Lb. bulgaricus* are lower when grown at 40°C than at 37°C (3 moles ATP per mole of glucose at 40°C and 2 at 37°C). This suggests that the coupling of growth to ATP production has 'slipped' at the higher temperature (Dirar & Collins, 1972). The surplus ATP may be hydrolysed by an ATPase, which may offer an advantage to the organism in providing energy for an uptake mechanism.

Molar growth yields have also been compared in *Str. cremoris* and *Str. lactis* subsp. *diacetylactis*; these organisms seem to be more efficient at coupling ATP synthesis to synthesis of cellular materials (Brown & Collins, 1977) although in complex media they may require less energy for maintenance (Stouthamer & Bettenhausen, 1975).

4.2. Associative Growth

The best documented example of associative growth is that of *Str. thermophilus* and *Lb. bulgaricus*. Pette & Lolkema (1950a) noted a rapid acid production from mixed cultures compared with single cultures and concluded that *Lb. bulgaricus* provided amino acids from the breakdown of milk proteins which stimulated growth of *Str. thermophilus* (Pette & Lolkema, 1950b). *Str. thermophilus* in return produced formic acid which stimulated *Lb. bulgaricus* (Galesloot *et al.*, 1968). Formic acid, however, is probably not the only stimulatory substance. Using the chemostat to study associative growth of these two organisms, Driessen *et al.* (1982) and Driessen (1981) have shown that CO_2 is also stimulatory to *Lb. bulgaricus*. The relationship between these two organisms is not an obligatory one so is proto-cooperative rather than symbiotic.

Mixed cultures are used for cheese and buttermilk manufacture. They are often dominated by *Str. cremoris* and *Str. lactis* which are responsible for lactic acid production; and in ripened products such as cheese, play an important role in the maturation process by contributing significant proteolytic and lipolytic activity. The aroma-forming bacteria used for buttermilk (*Leuconostoc* species, and *Str. lactis* subsp. *diacetylactis*) contribute small but important amounts of other volatiles. Production of diacetyl by the leuconostocs is favoured by low pH (Gilliland, 1972; Petterssen, 1975) and combination of these bacteria with *Str. cremoris* and/or *Str. lactis*, which lower pH quite rapidly, is therefore an advantage.

A mixed flora may not be stable because of cooperative effects alone. It is important that one population does not out-grow the other. Inhibitory effects of one population towards another may therefore be an important consideration. A stable microflora which is composed of many different types of organism, such as that of kefir, may be resistant to compositional change because all organisms are growing and metabolizing at less than their maximum capability. Kefir starter is composed of lactobacilli and streptococci whose optimum growth temperature is $\geq 30°C$, but kefir is made by fermenting milk at 22–25°C and ripening between 8–10°C. Under these conditions bacteria would not outgrow the yeast population and the yeast occupy only 5–10% of the flora (Kosikowski, 1977). Yeasts, however, may in turn be inhibited if the mixed flora contained propionate-producing propionibacteria, and may also be inhibited by diacetyl (Jay, 1982) produced by the streptococci.

5. CONCLUSIONS

Although the dairy lactic-acid bacteria require a wide variety of nutrients, ranging from metal ions to complex vitamins, detailed information on the mechanisms of nutrient utilization are confined to carbohydrates and proteins, reflecting the relative importance of these two categories of milk constituents to the growth of the organisms in fermented dairy products. Carbohydrate metabolism is particularly well documented, but the study of protein utilization (despite recent extensive literature) has not been sufficiently systematic to allow us to build up a complete picture of the interrelationships between proteolytic systems and transport systems. Nevertheless our present understanding of the physiology of the dairy lactic acid bacteria, however incomplete, is helpful in arriving at rational decisions as to the best starter strains for particular products, and the best conditions under which to culture them.

REFERENCES

ANDERS, R. F. & JAGO, G. R. (1970) *J. Dairy Res.* **37**, 445.
ANDREWS, K. J. & LIN, E. C. C. (1976) *Federation Proc.* **35**, 2185.
ARGYLE, P. J., MATHISON, G. E. & CHANDAN, R. C. (1976) *J. Appl. Bacteriol.* **41**, 175.

ASGHAR, S. S., LEVIN, E. & HAROLD, F. M. (1973) *J. Biol. Chem.* **248**, 5225.
BRACQUART, P. & LORIENT, D. (1979) *Milchwissenschaft* **34**, 676.
BREHENY, S., KANASAKI, M., HILLIER, A. J. & JAGO, G. R. (1975) *Australian J. Dairy Technol.* **30**, 145.
BROWN, W. V. & COLLINS, E. B. (1977) *Appl. Environ. Microbiol.* **33**, 38.
CHANDAN, R. C., ARGYLE, P. J. & MATHISON, G. E. (1982) *J. Dairy Sci.* **65**, 1408.
CHASSY, B. M. & THOMPSON, J. (1983) *J. Bacteriol.* **154**, 1195.
COGAN, T. M. (1981) *J. Dairy Res.* **48**, 489.
COGAN, T. M., O'DOWD, M. & MELLERICK, D. (1981) *Appl. Environ. Microbiol.* **41**, 1.
COWMAN, R. A. & SPECK, M. L. (1967) *Appl. Microbiol.* **15**, 851.
DAVIES, F. L. & GASSON, M. J. (1981) *J. Dairy Res.* **48**, 363.
DEMAIS, L. & VIANNA, A. L. (1979) *Ann. Rev. Biochem.* **48**, 275.
DESMAZEAUD, M. J. & HERMIER, J. H. (1972) *European J. Biochem.* **28**, 190.
DESMAZEAUD, M. J. & JUGE, M. (1976) *Le Lait* **56**, 241.
DIRAR, H. & COLLINS, E. B. (1972) *J. General Microbiol.* **73**, 233.
DRIESSEN, F. (1981) In *Mixed Culture Fermentations,* P. Bushell & J. Slater, Eds. Academic Press, London.
DRIESSEN, F. M., KINGMA, F. & STADHOUDERS, J. (1982) *Netherlands Milk Dairy J.* **36**, 135.
EGGIMAN, B. & BACHMANN, M. (1980) *Appl. Environ. Microbiol.* **40**, 876.
EL SODA, M. & DESMAZEAUD, M. J. (1981) *Agric. Biol. Chem.* **45**, 1693.
EL SODA, M. & DESMAZEAUD, M. J. (1982) *Canadian J. Microbiol.* **28**, 1181–8.
EL SODA, M., DESMAZEAUD, M. J. & BERGERE, J.-L. (1978a) *J. Dairy Res.* **45**, 445.
EL SODA, M., BERGERE, J.-L. & DESMAZEAUD, M. J. (1978b) *J. Dairy Res.* **45**, 519.
EL SODA, M., ZEYADA, N., DESMAZEAUD, M. J., MASHALY, R. & ISMAIL, A. (1982) *Sciences des Aliments* **2**, 261.
EXTERKATE, F. A. (1975) *Netherlands Milk Dairy J.* **29**, 303.
EXTERKATE, F. A. (1976) *Netherlands Milk Dairy J.* **30**, 95.
EXTERKATE, F. A. (1977) *J. Bacteriol.* **129**, 1281.
EXTERKATE, F. A. (1979a) *Arch. Microbiol.* **120**, 247.
EXTERKATE, F. A. (1979b) *J. Dairy Res.* **46**, 473.
EXTERKATE, F. A. (1982) *Irish J. Food Sci. Technol.* **6**, 205.
FOX, C. F. & KENNEDY, E. P. (1965) *Proc. Nat. Acad. Sci., USA* **54**, 891.
FRANKS, P. A., HALL, R. J. & LINKLATER, P. M. (1980) *Biotechnol. Bioeng.* **22**, 1465.
GALESLOOT, T. E., HASSING, F. & VERINGA, H. A. (1968) *Netherlands Milk Dairy J.* **22**, 50.
GARNIER, J., GAYE, P., MERCIER, J.-C. & ROBSON, B. (1980) *Biochimie* **62**, 231.
GASSON, M. J. (1983) *J. Bacteriol.* **154**, 1.
GILLILAND, S. E. (1972) *J. Dairy Sci.* **55**, 1028.
HAMILTON, W. A. (1975) *Advances Microbial Physiol.* **12**, 1.
HARVEY, R. J. & COLLINS, E. B. (1961) *J. Bacteriol.* **82**, 954.

HARVEY, R. J. & COLLINS, E. B. (1963) *J. Bacteriol.* **86**, 1301.

HEMME, D., NARDI, M. & WAHL, D. (1981a) *Le Lait* **61**, 1.

HEMME, D. H., SCHMAL, V. & AUCLAIR, J. E. (1981b) *J. Dairy Res.* **48**, 130.

HERMSDORF, C. L. & SIMMONDS, S. (1980) In *Microorganisms and Nitrogen Sources,* J. W. Payne, Ed. John Wiley & Sons, Chichester, p. 301.

JAY, J. M. (1982) *Appl. Environ. Microbiol.* **44**, 525.

JONAS, H. A., ANDERS, R. F. & JAGO, G. R. (1972) *J. Bacteriol.* **111**, 397.

KACZOROWSKI, G. T., ROBERTSON, D. E. & KABACK, H. R. (1979) *Biochemistry* **18**, 3697.

KASHKET, E. R. & WILSON, T. H. (1972) *Biochem. Biophys. Res. Communications* **49**, 615.

KEENAN, T. W. & LINDSAY, R. C. (1966) *J. Dairy Sci.* **50**, 1585.

KEMPLER, G. M. & MCKAY, L. L. (1980) *Appl. Environ. Microbiol.* **38**, 926.

KEMPLER, G. M. & MCKAY, L. L. (1981) *J. Dairy Sci.* **64**, 1527.

KONINGS, W. N. & ROBILLARD, G. T. (1982) *Proc. Nat. Acad. Sci., USA* **79**, 5480.

KOSIKOWSKI, F. V. (1977) In *Cheese and Fermented Milk Foods.* Edwards Brothers, Michigan.

LAUE, P. & MACDONALD, R. E. (1968a) *J. Biol. Chem.* **243**, 680.

LAUE, P. & MACDONALD, R. E. (1968b) *Biochim. Biophys. Acta* **165**, 410.

LAW, B. A. (1977) *J. Dairy Res.* **44**, 309.

LAW, B. A. (1978) *J. General Microbiol.* **105**, 113.

LAW, B. A. (1979) *J. Appl. Bacteriol.* **46**, 455.

LAW, B. A., & KOLSTAD, J. (1983) *Antonie van Leeuwenhoek J. Microbiol.* **49**, 209.

LAW, B. A., SEZGIN, E. & SHARPE, M. E. (1976) *J. Dairy Res.* **43**, 291.

LAWRENCE, R. C. & THOMAS, T. D. (1979) In *Soc. General Microbiol.* Symposium Series Vol. 29, A. T. Ball, D. C. Ellwood & C. Ratledge, Eds. Cambridge University Press, p. 187.

LAWRENCE, R. C., THOMAS, T. D. & TERZAGHI, B. E. (1976) *J. Dairy Res.* **43**, 141.

LE BLANC, D. J., CROW, V. L., LEE, L. N. & GARON, C. F. (1979) *J. Bacteriol.* **137**, 878.

LEES, G. J. & JAGO, G. R. (1976) *J. Dairy Res.* **43**, 75.

LEES, G. J. & JAGO, G. R. (1978a) *J. Dairy Sci.* **61**, 1205.

LEES, G. J. & JAGO, G. R. (1978b) *J. Dairy Sci.* **61**, 1216.

LEWIS, W. H. P. & HARRIS, H. (1967) *Nature, London* **215**, 351.

MALONEY, P. C. (1977) *J. Bacteriol.* **132**, 564.

MALONEY, P. C. & WILSON, T. H. (1975) *J. Membrane Biol.* **25**, 285.

MARSHALL, V. M., COLE, W. M. & MABBITT, L. A. (1982) *J. Dairy Res.* **49**, 147.

MASON, P. W., CARBONE, D. P., CUSHMAN, R. A. & WAGGONER, A. S. (1981) *J. Biol. Chem.* **256**, 1861.

MCKAY, L. L. & BALDWIN, L. A. (1974) *Appl. Microbiol.* **28**, 342.

MCKAY, L. L., WALTER, L. A., SANDINE, W. E. & ELLIKER, P. R. (1969) *J. Bacteriol.* **99**, 603.

MCKAY, L. L., MILLER, A., SANDINE, W. E. & ELLIKER, P. R. (1970) *J. Bacteriol.* **102**, 804.

MELLERICK, D. & COGAN, T. M. (1981) *J. Dairy Res.* **48**, 497.

MILLS, O. E. & THOMAS, T. D. (1978) *New Zealand J. Dairy Sci. Technol.* **13**, 209.

MILLS, O. E. & THOMAS, T. D. (1981) *New Zealand J. Dairy Sci. Technol.* **16**, 43.

MITCHELL, P. (1963) *Biochem. Soc. Symp.* **22**, 142.

MITCHELL, P. (1970) *Symp. Soc. General Microbiol.* **20**, 121.

MITCHELL, P. (1973) *J. Bioenergetics* **4**, 65.

MORISHITA, T., DEGUCHI, Y., YAJIMA, M., SAKURAI, T. & YURA, T. (1981) *J. Bacteriol.* **148**, 64–71.

MOU, L., SULLIVAN, J. J. & JAGO, G. R. (1975) *J. Dairy Res.* **42**, 147.

NIVEN, D. F. & HAMILTON, W. A. (1973) *FEBS Lett.* **37**, 244.

NIVEN, D. F., JEACOCKE, R. E. & HAMILTON, W. A. (1973) *FEBS Lett.* **29**, 248.

O'LEARY, V. S. & WOYCHICK, J. H. (1976) *Appl. Environ. Microbiol.* **32**, 89.

OLIVER, W. H. (1953) *J. General Microbiol.* **8**, 38.

OLSON, H. C. & QUTUB, A. H. (1970) *Cultured Dairy Prod. J.* **5**, 12.

OTTO, R., LAGEVEEN, R. G., VELDKAMP, H. & KONINGS, W. N. (1982) *J. Bacteriol.* **149**, 733.

OTTO, R., TENBRINK, B., VELDKAMP, H. & KONINGS, W. N. (1983) *FEMS Microbiol. Lett.* **16**, 69.

PARK, Y. H. & McKAY, L. L. (1982) *J. Bacteriol.* **149**, 420.

PAYNE, J. W. (1980) In *Microorganisms and Nitrogen Sources,* J. W. Payne, Ed. John Wiley & Sons, Chichester, p. 211.

PEARCE, L. E., BRICE, S. A. & CRAWFORD, A. M. (1973) *New Zealand J. Dairy Sci. Technol.* **8**, 17.

PETERKOVSKY, A. (1981) *Advances Cyclic Nucleotide Res.* **14**, 215.

PETTE, J. W. & LOLKEMA, H. (1950a) *Netherlands Milk Dairy J.* **4**, 209.

PETTE, J. W. & LOLKEMA, H. (1950b) *Netherlands Milk Dairy J.* **4**, 261.

PETTERSSEN, H. E. (1975) *Appl. Microbiol.* **29**, 133.

POLLOCK, M. R. (1962) In *The Bacteria,* Vol. IV, I. C. Gunsalus & R. Y. Stanier, Eds. Academic Press, New York, p. 121.

PREMI, L., SANDINE, W. E. & ELLIKER, P. R. (1972) *Appl. Microbiol.* **24**, 51.

RACLIFF, T. L. & TALBART, D. E. (1981) *J. Dairy Sci.* **64**, 391.

RACLIFF, T. L., STINSON, R. S. & TALBURT, D. E. (1980) *Canadian J. Microbiol.* **26**, 58.

REDDY, M. S., WILLIAMS, F. D. & REINBOLD, G. W. (1973) *J. Dairy Sci.* **56**, 634.

REIDER, E., WAGNER, E. F. & SCHWEIGER, M. (1979) *Proc. Nat. Acad. Sci., USA* **76**, 5529.

REITER, B. & ORAM, J. D. (1962) *J. Dairy Res.* **29**, 63.

RICE, G. H., STEWART, F. H. C., HILLIER, A. J. & JAGO, G. R. (1978) *J. Dairy Res.* **45**, 93.

ROBILLARD, G. T. (1982) *Molecular Cellular Biochem.* **46**, 3.

ROBILLARD, G. T. & KONINGS, W. N. (1981) *Biochem.* **20**, 5025.

ROSS, G. D. (1980) *Australian J. Dairy Technol.* **35**, 147.

SAHYOUN, N. & DURR, I. F. (1972) *J. Bacteriol.* **112**, 421.

SANDINE, W. E. & ELLIKER, P. R. (1970) *J. Agric. Food Chem.* **18**, 557.

SANDINE, W. E., RADICH, P. C. & ELLIKER, P. R. (1972) *J. Milk Food Technol.* **35**, 176.

SEITZ, E. W., SANDINE, W. E., ELLIKER, P. R. & DAY, E. A. (1963) *Canadian J. Microbiol.* **9**, 431.

SEARLES, M. A., ARGYLE, P. J., CHANDAN, R. C. & GORDON, J. F. (1970) *Proc. XVIII Int. Dairy Congr., Sydney* **IE**, 111.

SELBY-SMITH, J., HILLIER, A. J., LEES, G. J. & JAGO, G. R. (1975) *J. Dairy Res.* **42**, 123.

SHANKAR, P. A. (1977) Thesis, University of Reading, UK.

SHANKAR, P. A. & DAVIES, F. L. (1978) *Proc. XX Int. Dairy Congr., Paris,* 467.

SHIRES, T. K., PITOT, H. C. & KAUFFMAN, S. A. (1974) *Biomembranes* **5**, 81.

SØRHAUG, T. & SOLBERG, P. (1973) *Appl. Microbiol.* **25**, 388.

SØRHAUG, T. & KOLSTAD, J. (1981) *Netherlands Milk Dairy J.* **35**, 338.

STOUTHAMER, A. H. & BETTENHAUSEN, G. W. (1975) *Arch. Microbiol.* **102**, 187.

TANIGUCHI, K., NAGAO, A. & TSUGO, T. (1965) *Japanese J. Zootechnic. Sci.* **36**, 376.

TERZAGHI, B. E. & SANDINE, W. E. (1975) *Appl. Microbiol.* **29**, 807.

THOMAS, T. D. (1975) *Biochem. Biophys. Res. Communications* **63**, 1035.

THOMAS, T. D. (1976a) *J. Bacteriol.* **125**, 1240.

THOMAS, T. D. (1976b) *Appl. Environ. Microbiol.* **32**, 474.

THOMAS, T. D. (1979) *New Zealand J. Dairy Sci. Technol.* **14**, 12.

THOMAS, T. D. & CROW, V. L. (1983) *FEMS Microbiol. Lett.* **17**, 13.

THOMAS, T. D., JARVIS, B. D. W. & SKIPPER, N. A. (1974) *J. Bacteriol.* **118**, 329.

THOMAS, T. D., TURNER, K. W. & CROW, V. L. (1980) *J. Bacteriol.* **144**, 672.

THOMPSON, J. (1976) *J. Bacteriol.* **127**, 719.

THOMPSON, J. (1978) *J. Bacteriol.* **136**, 465.

THOMPSON, J. (1980) *J. Bacteriol.* **144**, 683.

THOMPSON, J. & THOMAS, T. D. (1977) *J. Bacteriol.* **130**, 583.

THOMPSON, J., TURNER, K. W. & THOMAS, T. D. (1978) *J. Bacteriol.* **133**, 1163.

TINSON, W., HILLIER, A. J. & JAGO, G. R. (1982a) *Australian J. Dairy Technol.* **37**, 8.

TINSON, W., BROOME, M. C., HILLIER, A. J. & JAGO, G. R. (1982b) *Australian J. Dairy Technol.* **37**, 14.

ULLMANN, A. & DANCHIN, A. (1982) *Advances Cyclic Nucleotide Res.* **15**, 1–54.

VESCOVA, M. & BOTTAZZI, V. (1979) *Scienza e Technica Lattiero Casearia* **30**, 434.

VIGNAIS, P. V. (1976) *Biochim. Biophys. Acta* **456**, 1.

WEST, I. C. & MITCHELL, P. (1973) *Biochem. J.* **132**, 587.

WILLIAMSON, W. T., TOVE, S. B. & SPECK, M. L. (1964) *J. Bacteriol.* **87**, 49.

Chapter 4

The Genetics of Dairy Lactic-acid Bacteria

Michael J. Gasson and F. Lyndon Davies

National Institute for Research in Dairying, Shinfield, Reading, UK

1. INTRODUCTION

Interest in the genetics of lactic-acid bacteria is rapidly expanding. This is partly because the genetic technology now being developed will open the way for industrial strain improvement programmes in the dairy and food industries, as well as offering the possibility of creating completely novel strains for new commercial uses. In addition, the genetic phenomena so far discovered promise a wealth of scientific interest for the academic geneticist and molecular biologist. This is a young field of research, and it will be clear that there are many gaps in our understanding and technical capabilities. However, with more scientists taking an interest in these little-studied bacteria, our knowledge is steadily growing. The development of new industrial strains by genetic manipulation remains a goal for the future, but the prospect has led to a range of speculation from the wildly optimistic to the sceptical and has also raised questions of safety and acceptability in what many regard as a conservative and traditional industry. It is our intention in this chapter to provide an up-to-date review of the genetic makeup and the gene transfer processes known to exist in lactic-acid bacteria and also to present our views on the potential applications for this exciting new area of biotechnology.

2. ARRANGEMENT OF THE GENETIC MATERIAL

The genes of bacteria are arranged in two ways; as a major long DNA molecule, the bacterial chromosome, and as additional, smaller non-essential units of DNA, the autonomously replicating plasmids. The genetics of lactic-acid bacteria largely concerns their plasmids and their bacteriophages, partly because they are important, but also because they are more accessible than the chromosome for molecular genetic analysis. It has long been recognized that certain properties of starter strains, vital for successful dairy fermentations, are unstable and loss of plasmid molecules goes some way towards accounting for this phenomenon. Bacteriophage attack is one of the most serious recurring problems of milk fermentation and hence a detailed discussion of bacterial viruses is included as a separate chapter (Chapter 5). Here we consider the plasmids.

2.1. Detection and Abundance of Plasmid DNA

The discovery of extrachromosomal DNA in lactic streptococci followed speculation that loss of lactose-utilizing ability in *Str. lactis* might be caused by plasmid curing (McKay *et al.*, 1972). Initially, plasmid DNA was resolved as a satellite peak after equilibrium density gradient centrifugation of radioactively labelled DNA in caesium chloride and ethidium bromide (Cords *et al.*, 1974). Electron microscopy of this DNA revealed circular plasmid molecules of various sizes which could be estimated by contour length measurement (Cords *et al.*, 1974; McKay & Baldwin, 1975; Anderson & McKay, 1977). Now, the more rapid and convenient technique of agarose gel electrophoresis (Meyers *et al.*, 1976) is generally used to detect the plasmids of lactic-acid bacteria (Klaenhammer *et al.*, 1978). Before plasmid detection methods can be used, cells need to be lysed and a substantial quantity of chromosomal DNA removed. Various protocols for achieving this have been reported and they fall into two groups; simple cleared lysate methods in which the cell membrane with attached chromosomal DNA is precipitated and removed by centrifugation (Klaenhammer *et al.*, 1978; Gasson & Davies, 1980b) and methods involving the denaturation of chromosomal DNA by high pH (Walsh & McKay, 1981; Anderson & McKay, 1983a). The latter techniques are preferable where plasmids with molecular weights in excess of 30 megadaltons (Mdal) are to be reliably detected. In addition, streamlined protocols have been developed for rapidly screening many strains for their plasmid content (LeBlanc & Lee, 1979; Gasson, 1983; Anderson & McKay, 1983a; Yu *et al.*, 1982).

Strains of *Str. lactis* and *Str. cremoris* contain remarkably large and complex plasmid complements (Fig. 1). The observed number of molecules ranges from 2 to 11 but between 4 and 7 is more typical (McKay, 1983). The plasmid profiles provide a useful 'fingerprint' which can be exploited in a simple way to investigate strain identities and interrelationships (Davies *et al.*, 1981). In contrast, strains of *Str. thermophilus* are often plasmid-free or contain small numbers of

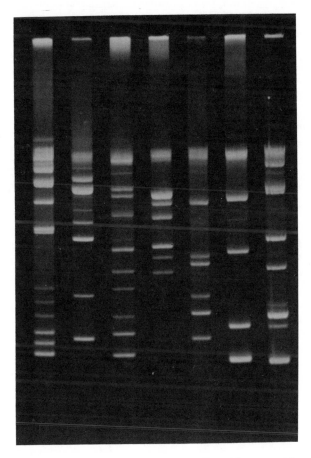

Fig. 1. Plasmid profiles in lactic streptococci. Plasmids present in a variety of different strains have been separated by agarose gel electrophoresis.

plasmids (Somkuti & Steinberg, 1981; Davies & Underwood, unpublished). The plasmid complements of lactobacilli were found by Vescovo *et al.* (1981) to be variable, often containing none or only one plasmid, although some strains of *Lb. acidophilus* had many plasmids. Pediococci examined by Gonzales & Kunka (1983) contained two, one or no plasmids, and a *Leuconostoc* examined by Moonan & Warner (personal communication) carried two plasmids.

2.2. Plasmid Curing and Phenotype Assignment

Evidence that a phenotype is plasmid controlled includes observation of its spontaneous loss, detecting an increased frequency of curing after growth in the presence of acriflavine, ethidium bromide or at elevated temperatures, and genetic transfer of the character between strains. Ultimately, however, it is desirable to assign the phenotype to a particular molecule from a strain's plasmid complement and in many instances this is simply achieved by observing a consistent correlation between phenotype loss and curing of a particular plasmid molecule (Fig. 2). In some cases, for instance the lactose plasmids of *Str. lactis* 712 and *Str. lactis* subsp. *diacetylactis* 176, such clear evidence is not readily obtained. In this situation, elimination of plasmid molecules by curing has proved to be of value. In *Str. lactis* 712 random plasmid elimination was observed after the regeneration of protoplasts and was exploited to sequentially cure plasmids and isolate derivative strains carrying individual molecules as well as a plasmid-free strain (Fig. 3). Observation of the phenotype of these 'protoplast cured' strains clearly identified a lactose plasmid molecule, even though this was not possible by conventional phenotype curing and subsequent plasmid profile analysis (Gasson, 1983). The complex plasmid complements of lactic streptococci can also pose problems for analysis of the physical changes brought about by genetic transfer of plasmid-encoded characters. Furthermore, they complicate the purification of an individual plasmid for detailed molecular analysis. For these reasons the isolation of single plasmid-carrying strains and plasmid-free strains is of considerable value. As well as the cited example of *Str. lactis* 712, plasmid-free derivatives have been obtained from *Str. lactis* C2 after mutagenesis with N-methyl-N'-nitro-N-nitrosoguanidine (McKay *et al.*, 1976), from *Str. lactis* TL1387 by 'protoplast curing' (A. Chopin, personal communication); from *Str. lactis* subsp. *diacetylactis* 176 after acriflavine induced curing (Gasson, 1984b) and from *Str. lactis* subsp. *diacetylactis* BU2 (Teuber & Lembke, 1983) and *Str. cremoris* 1200 (Davies, unpublished) by growth at elevated temperatures.

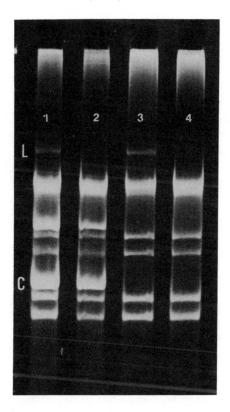

Fig. 2. Assignment of plasmid phenotypes by curing. Plasmid profiles of *Str. lactis* subsp. *diacetylactis* NCDO1008 (1) and protoplast cured derivatives with Lac^-Cit^+ (2), Lac^+Cit^- (3) and Lac^-Cit^- (4) phenotypes are shown. Plasmids controlling lactose utilization (L) and citrate utilization (C) are clearly identified (Gasson, 1983).

2.3. Lactose and Proteinase Plasmids

Phenotypic, genetic and physical evidence has been used to show that in many lactic streptococci lactose-fermenting ability is a plasmid-encoded property. Spontaneous loss of this phenotype was noted by workers in the 1930s (e.g., Sherman & Hussong, 1937) and McKay *et al.* (1972) showed that frequencies of curing were increased by growth in the presence of acridine dyes or at elevated temperatures. In many strains of *Str. lactis* and *Str. lactis* subsp. *diacetylactis* lactose genes have been assigned to particular molecules and the available information is summarized in Table 1. Although curing experiments identified

Michael J. Gasson and F. Lyndon Davies

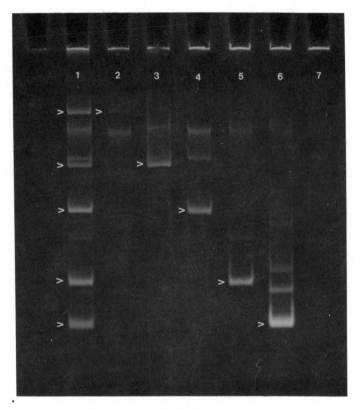

Fig. 3. Creation of plasmid-free and single plasmid-carrying derivatives of *Str. lactis* 712. The *Str. lactis* 712 plasmid profile (1) is followed by: MG1299 with just the lactose proteinase plasmid (2); MG3197 carrying just the aggregation plasmid pAG712 (3); MG1261 (4), MG1362 (5), and MG1365 (6) carrying individual cryptic plasmids; and MG1363 (7) a plasmid-free strain. The derivatives were obtained by sequential rounds of protoplast curing (Gasson, 1983). The CCC form of the plasmid DNA is arrowed in each strain.

a 36 Mdal lactose plasmid in *Str. cremoris* B$_1$, curing and physical evidence for lactose plasmids in many other strains of this species was not obtained (Larsen & McKay, 1978; Davies, unpublished). Even so, transfer of lactose genes was correlated with transfer of a plasmid molecule when several *Str. cremoris* strains were used as donors in conjugal mating experiments with *Str. lactis* recipients. This has led Snook & McKay (1981) to speculate that *Str. cremoris* strains may carry unusually stable lactose plasmids. Against this is the observation that a plasmid-free derivative of *Str. cremoris* 1200 retained lactose fermenting ability (Davies, unpublished). Hence the status of the lactose phenotype in *Str. cremoris* varies, some strains carry lactose plasmids, some seem only to have chromosomal lactose genes, and there is always the possibility of lactose plasmids stabilized by their integration into the chromosome. Lactose plasmids have also been identified in a number of *Lactobacillus* strains and again these are listed in Table 1.

In some strains efforts have been made to determine precisely which enzymes of the lactose catabolic pathway are plasmid-controlled. Two lactose specific components of the PEP-dependent phosphotransferase

TABLE 1
Plasmid-encoded Lactic and Proteinase Genes (Sizes of plasmids reported to encode lactose and/or proteinase genes are given in megadaltons)

Species	Strain	Lac	Prt	Lac/Prt	Reference
Str. lactis	C2	30	18/12·5	—	Klaenhammer *et al.* (1978)
	ML3	33	—	33	Walsh & McKay (1982)
					Kuhl *et al.* (1979)
	712	—	—	33	Gasson (1982, 1983)
	C10	—	—	40	Kuhl *et al.* (1979)
	M18	—	—	45	Kuhl *et al.* (1979)
	DR1251	32	—	—	LeBlanc *et al.* (1979)
	H1	33	—	—	Crow *et al.* (1983)
Str. lactis subsp.	18-16	41	—	—	Kempler & McKay (1979b)
diacetylactis	DRC3	52	—	—	McKay *et al.* (1980)
	11007	32	—	—	McKay *et al.* (1980)
	DRC1	—	—	31	Kempler & McKay (1979a)
Str. cremoris	B1	37	—	—	Klaenhammer *et al.* (1978)
	HP	—	9	—	Larsen & McKay (1978)
	C3	34/27	—	—	Snook & McKay (1981)
	R1	34	—	—	Snook & McKay (1981)
	EB7	42	—	—	Snook & McKay (1981)
	Wg2	—	16	—	Otto *et al.* (1982)
	SK11	—	48	—	Davies (unpublished)
	1200	—	14	—	Davies (unpublished)
Lb. helveticus subsp. *jugurti*	536·2	8·2	—	—	Smiley & Fryder (1978)
Lb. casei	DR1002	23	—	—	Chassy *et al.* (1978)
	64H	21	—	—	Lee *et al.* (1982)

(PTS) system for sugar uptake (enzyme II and factor III), as well as the lactose splitting enzyme phospho-β-galactosidase, were found to be determined by plasmid-encoded genes in *Str. lactis* C2 and in *Str. cremoris* B1 (Anderson & McKay, 1977; McKay, 1983). Transformation into lactose-negative mutants of *Str. sanguis* was used by St. Martin *et al.* (1982) to show that the *Str. lactis* 11454 lactose plasmid carried genes for the lactose phosphotransferase system and phospho-β-galactosidase. In addition to these functions, Crow *et al.* (1983) found that genes for three enzymes of the D-tagatose-6-phosphate pathway (galactose-6-phosphate isomerase, D-tagatose-6-phosphate kinase and tagatose-1,6-diphosphate aldolase) were linked to the lactose plasmids of *Str. lactis* strains C10, H1 and 133. The lactose plasmid of *Lb. casei* 64H has also been shown to encode genes for the lactose phosphotransferase system and for phospho-β-galactosidase (Lee *et al.*, 1982). Utilization of galactose is also affected by loss of the lactose plasmid. It has been known for some time that lactose-negative variants of *Str. lactis* have slower growth rates on galactose (McKay *et al.*, 1970; Demko *et al.*, 1972; Cords & McKay, 1974; LeBlanc *et al.*, 1979). This has been explained by the existence of two pathways for galactose catabolism. One involves uptake by the PEP-dependent phosphotransferase system and is controlled by the lactose plasmid, the other probably uses a chromosomally encoded APT permease system. Although it was originally thought that the lactose-specific PTS system also accounted for plasmid-controlled galactose uptake, a separate galactose PTS system has recently been discovered and shown to be encoded by the lactose plasmid of *Str. lactis* C2 (Thompson, 1980; Park & McKay, 1982).

Because lactic streptococci are nutritionally fastidious and significant amounts of free amino acids or small peptides are not found in milk, efficient starter growth and acid production depends on casein breakdown by proteinases bound to the starter cell wall. Loss of this proteolytic ability, resulting in slow acid production, has been recognized as a starter problem for many years (e.g., Harriman & Hammer, 1931; Garvie & Mabbitt, 1956). The suggestion that spontaneous plasmid curing may cause such slow variants was made by Pearce (1970) and physical evidence for this was first obtained for *Str. lactis* C2 by McKay & Baldwin (1975) and Klaenhammer *et al.* (1978). Proteinase plasmids have now been identified in a number of strains of *Str. cremoris,* and details of these are included in Table 1.

The identification of proteinase (*Prt*) plasmids in *Str. lactis* has caused considerable confusion mainly because the character is incompletely genetically linked to the lactose determinant (*Lac*). In *Str. lactis* C2, ML3, 712, C10, M18 and *Str. lactis* subsp. *diacetylactis* DRC1, simultaneous curing of both characters was correlated with loss of a large plasmid ranging in size from 30 to 45 Mdal (see Table 1). In contrast, proteinase-negative variants of *Str. lactis* C2 which retained lactose genes lost 12·5 Mdal and 18 Mdal plasmids but appeared to retain the large 30 Mdal molecule (McKay & Baldwin, 1974; Klaenhammer *et al.*, 1978). Gene transfer experiments caused further complications in that lactose gene transfer was sometimes, but not always, accompanied by co-transfer of proteinase genes, both in transduction experiments (McKay *et al.*, 1976; Davies & Gasson, 1981), and in conjugation experiments (Gasson & Davies, 1980b; Gasson, 1983). The recent discovery of high frequency deletion formation (Gasson, 1982, 1983) and the observation of very high frequency curing of a cryptic plasmid (Davies & Gasson, 1981; Gasson, 1983) are especially relevant in clarifying these rather contradictory reports. First, curing of a 9 Mdal plasmid from *Str. lactis* 712 occurs very readily and often such cured strains are lactose- and proteinase-negative, but not always. The 9 Mdal plasmid does not control lactose or proteinase genes but rather an aggregation phenotype (Gasson & Elkins, unpublished). From this it is clear that loss of a plasmid in association with a change in phenotype is only relevant if such a correlation is consistent. Second, the lactose and proteinase plasmid of *Str. lactis* 712 is subject to very frequent deletion formation. Deletions can generate Lac^-Prt^-, Lac^-Prt^+ and Lac^+Prt^- phenotypes depending where on the molecule they occur (Gasson, 1983). Furthermore, deletion of 10% or less of the plasmid DNA would not be detected when plasmid profiles were examined by agarose gel electrophoresis. Transduction of pLP712 demands deletion formation in order to accommodate plasmid DNA in the bacteriophage head (see Section 3.1) and again depending on the deletion position, Lac^+Prt^+ or Lac^+Prt^- transductants result. Lactose genes have also been conjugally transferred from the related *Str. lactis* strains 712, ML3 and C2 into a plasmid-free *Str. lactis* 712 recipient. For all three donors both Lac^+Prt^+ and Lac^+Prt^- progeny were isolated and the plasmids present were compared by restriction endonuclease digestion. Plasmids in the Lac^+Prt^+ progeny appeared identical regardless of the donor strain used. The Lac^+Prt^- progeny carried a similar plasmid but with a deletion in a region known to encode proteinase genes (Gasson, unpublished). It seems likely that *Str. lactis* strains commonly carry large plasmids which encode both lactose and proteinase genes, but that they may undergo deletion and break the genetic linkage between these two phenotypes.

The lactose and proteinase plasmid pLP712 has been extensively mapped by both restriction endonuclease digestion and deletion mapping and is further described in Section 4.

2.4. Citrate Plasmids

As with lactose and proteinase genes, it has been known for over half a century that the ability of *Str. lactis* subsp. *diacetylactis* strains to ferment citrate (*Cit*) is an unstable property. Kempler & McKay (1979a) used acridine orange curing to isolate *Cit⁻* variants and strains 18-16 and DRC1. These *Cit⁻* derivatives had lost the ability to transport citrate but retained citratase activity and the change was correlated with curing of a 5·5 Mdal plasmid. It is now well established that a similar 5·5 Mdal citrate plasmid is present in many *Str. lactis* subsp. *diacetylactis* strains. Kempler & McKay (1981) used restriction endonuclease digestion to compare the citrate plasmids from *Str. lactis* subsp. *diacetylactis* strains (DRC1, DRC3, 11007, 18-16 and WM4 and found them to be identical. The citrate plasmid from *Str. lactis* subsp. *diacetylactis* 176 has been extensively restriction mapped (Gasson, 1984b). It also appears identical to other citrate plasmids and the current restriction endonuclease map is reproduced in Fig. 4.

2.5. Plasmids Controlling Antagonistic Properties

A variety of antagonistic substances are known to be produced by lactic streptococci and in some cases evidence for plasmid control has been reported.

Certain strains of *Str. lactis* produce a polypeptide antibiotic, nisin, which has a broad spectrum activity against gram-positive bacteria as well as their spores and is used as a food preservative. Kozak *et al.* (1974) were able to isolate spontaneous, stable nisin-negative variants from several nisin-producing strains and observed increased curing frequencies after growth with proflavine, ethidium bromide, acridine or at elevated temperatures. Attempts to satisfactorily correlate nisin-producing ability with plasmid DNA were unsuccessful (Fuchs *et al.*, 1975; Davies & Gasson, 1981). Davey & Pearce (1982) isolated a plasmid-free derivative of *Str. lactis* H1 but found the strain's ability to produce nisin was unaffected. In *Str. lactis* 11454, a 28 Mdal plasmid was reported to control both sucrose-fermenting ability and nisin

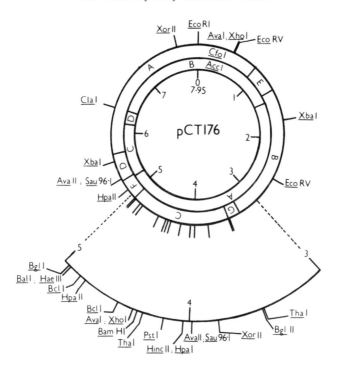

Fig. 4. Restriction endonuclease map of the citrate plasmid pCT176. The map is taken from a report of the molecular characterization and cloning of pCT176 (Gasson, 1984b).

production (LeBlanc *et al.*, 1980). This linkage of sucrose and nisin genes has recently been shown to be a general phenomenon (Gasson, 1984a; Hart, Davies & Gasson, unpublished). Sucrose fermenting ability was conjugally transferred from 8 different nisin-producing donor strains into a plasmid-free derivative of the nisin-sensitive, non-nisin-producing strain *Str. lactis* 712. All progeny from these experiments had also acquired nisin resistance and the ability to produce a nisin-like inhibitor. These properties were simultaneously cured and transferable from the progeny strains suggesting the existence of a transmissible sucrose and nisin plasmid (pSN) (Gasson, 1984a). Good physical evidence for such pSN plasmids has not however been obtained. Plasmid DNA was detected in the *Str. lactis* 712 progeny but it proved labile and results were difficult to repeat (Gasson, 1984a). In a related study (Hart, Davies & Gasson, unpublished), the plasmid profiles of sucrose- and nisin-cured derivatives isolated from six different nisin producer strains were carefully analysed. Evidence for loss of a pSN plasmid molecule was not obtained. Hence, there is clear genetic evidence of plasmid-encoded nisin and sucrose genes but such a plasmid molecule has yet to be convincingly identified.

A similar antibiotic, diplococcin, is produced by some *Str. cremoris* strains. Davey & Pearce (1982) isolated non-revertible heat-cured derivatives from several strains and transferred diplococcin-producing ability to *Str. lactis* ML3 and C10. Despite this genetic evidence for plasmid control, physical evidence for a diplococcin plasmid was not obtained.

Many strains of group N streptococci are capable of producing bacteriocin-like substances and bacteriocin-negative variants have been obtained by standard plasmid curing techniques (Dobrzanski *et al.*, 1982). In one strain, *Str. lactis* subsp. *diacetylactis* WM4, phenotypic, genetic and physical data have been used to clearly identify an 88 Mdal transmissible bacteriocin plasmid (McKay & Baldwin, 1982; Scherwitz *et al.*, 1983). Neve *et al.* (1984) used curing and conjugation experiments to identify bacteriocin plasmids in *Str. cremoris* strains 4G6 and 9B4 and in *Str. lactis* subsp. *diacetylactis* 6F7. Restriction endonuclease analysis was used to show that 40 Mdal bacteriocin plasmids in the *Str. cremoris* strains were virtually identical and that they formed a part of the larger 75 Mdal bacteriocin plasmid identified in *Str. lactis* subsp. *diacetylactis* 6F7. In addition, the same authors used Southern hybridization to reveal DNA homology between these transmissible bacteriocin plasmids and non-transmissible but curable bacteriocin plasmids found in *Str. cremoris* strains AC1 and 3A6.

2.6. Additional Plasmid-encoded Phenotypes

Two plasmid-encoded mechanisms of bacteriophage resistance, restriction modification and adsorption inhibition, have been identified and these are fully described in Sections 2.3 and 2.4 of Chapter 5.

Physical evidence for a 23 Mdal plasmid encoding glucose and mannose genes in *Str. lactis* 354-07, and a 3·6 Mdal xylose plasmid in *Str. lactis* DR1253 has been reported (LeBlanc *et al.*, 1980).

Dobrzanski *et al.* (1982) were able to transform a kanamycin resistance plasmid from *Str. lactis* 71 into *Bacillus subtilis*, but this is the only report of inherent drug resistance plasmids in lactic streptococci. Inorganic ion resistance or sensitivity have been identified as plasmid-encoded properties. The lactose plasmid of *Str. lactis* C2 was reported to control resistance to arsenate, arsenite and chromate and sensitivity to copper (Efstathiou & McKay, 1977) and a 6·1 Mdal plasmid from *Str. cremoris* Wg2 may control copper resistance (Otto *et al.*, 1982). Other traits that may be plasmid-controlled, but for which data have not been reported, include resistance to the lactoperoxidase–thiocyanate–hydrogen peroxide system, slime formation by *Str. cremoris* strains used for villi manufacture and arginine hydrolysis by *Str. lactis* subsp. *diacetylactis* strains (McKay, 1983).

3. GENE TRANSFER PROCESSES

A variety of different gene transfer processes have been discovered, exploited and analysed in lactic acid bacteria relevant to the dairy industry. In this section the various processes are described and their value for further developing the applied and academic genetics of these bacteria is discussed.

3.1. Transduction
The first detailed account of gene transfer in the lactic streptococci is of bacteriophage-mediated transduction in *Str. lactis* C2 (McKay *et al.*, 1973, 1976; McKay & Baldwin, 1974). The earlier literature reports transduction of streptomycin resistance in *Str. lactis* C2 (Allen *et al.*, 1963) and tryptophan independence in *Str. lactis* subsp. *diacetylactis* 18-16 (Sandine *et al.*, 1962). Although these early experiments involved virulent bacteriophages, McKay's work with *Str. lactis* C2 and all subsequent reports of gene transfer by transduction have involved temperate bacteriophages.

In *Str. lactis* C2 transduction of chromosomal maltose and mannose markers and plasmid-encoded lactose genes were detected (McKay *et al.*, 1973). In the latter case non-selected transfer of proteinase genes also occurred in approximately 50% of the lactose transductants selected (McKay & Baldwin, 1974). When transduction experiments were repeated using bacteriophages induced from a lactose transductant rather than from the parental *Str. lactis* C2 strain, the frequency of lactose gene transfer was found to be substantially increased (McKay *et al.*, 1976). This situation is well known in other transduction systems and is referred to as high frequency transduction (HFT). An essentially similar transduction process has been described in the related *Str. lactis* strain 712. In addition to the lactose plasmid, a 17 Mdal erythromycin resistance plasmid introduced from *Str. faecalis* (see Section 3.2) was transduced, but in this case the HFT phenomenon did not occur (Davies & Gasson, 1981). The molecular consequences of lactose plasmid transduction and an explanation for the HFT phenomenon in *Str. lactis* 712 have been determined and these are discussed in Section 4.

Recently a temperate bacteriophage induced from *Str. cremoris* C3 was shown to transduce lactose genes into *Str. lactis* ML3 and into a plasmid-free derivative of *Str. lactis* C2 (Snook *et al.*, 1981).

The value of transduction as a gene transfer process is limited, because only host strains which have the relevant bacteriophage receptors can be used and because the head capacity of the phage limits the amount of DNA that can be transferred. This latter fact does, however, make transduction very useful for making fine structural changes in the genetic material. Transduction of the lactose plasmid causes deletions to be generated because the whole molecule is too large to be accommodated in the phage head (McKay *et al.*, 1976; Gasson, 1983). These deletions are very useful for mapping studies in which plasmid-encoded genes need to be located on a restriction endonuclease cleavage map (Gasson, 1983, see Section 4).

In addition, transduction sometimes causes lactose and proteinase genes to become integrated into the bacterial chromosome with two important consequences; the genes become stabilized by their new location and the levels of gene expression are reduced (McKay & Baldwin, 1978; Snook *et al.*, 1981). *Str. lactis* C2 derivatives with chromosomally integrated proteinase genes are less proteolytic and produce a better cheese with reduced levels of bitter peptide (McKay & Baldwin, 1978).

3.2. Conjugation

Gene transfer by mating processes which demand cell to cell contact has been widely reported in lactic streptococci. First reports were of conjugal transfer of plasmid-encoded lactose genes between marked strains of *Str. lactis* 712 (Gasson & Davies, 1979, 1980b) and from *Str. lactis* subsp. *diacetylactis* 18-16 into a plasmid-free derivative of *Str. lactis* C2 (Kempler & McKay, 1979b). Subsequently lactose plasmid transfer has been detected into this same *Str. lactis* C2 derivative from *Str. lactis* ML3 and *Str. lactis* C2, from *Str. cremoris* C3 and from *Str. lactis* subsp. *diacetylactis* strains DRC3, 11007 and WM4; into *Str. lactis*

ML3 derivatives from *Str. cremoris* strains R1, EB7, 28 and C3; and also into a *Str. cremoris* B₁ derivative from *Str. cremoris* C3 (McKay *et al.*, 1980; Snook & McKay, 1981). A lactose plasmid from *Lb. casei* 64H has also been found to be capable of conjugal transfer (Lee *et al.*, 1982). Although these transfer processes were initially detected as relatively low frequency events, high frequency donors have been isolated amongst the progeny from some mating experiments. These variant strains, which were first discovered in *Str. lactis* 712 (Gasson & Davies, 1979, 1980b), exhibit a novel cell aggregation phenomenon which causes a striking change both in colony morphology and in the appearance of broth cultures. Similar, aggregating progeny strains were found after lactose plasmid transfer from *Str. lactis* ML3 into a plasmid-free derivative of *Str. lactis* C2 (Walsh & McKay, 1981). As well as this natural lactose plasmid of *Str. lactis* ML3, lactose plasmids introduced into this host from *Str. cremoris* strains C3 and Z8 generated aggregating progeny (Snook & McKay, 1981; Snook *et al.*, 1981). It is now well established that this acquisition of high frequency donor ability together with the cell aggregation phenotype involves enlargement of the lactose plasmids by insertion of novel DNA (Walsh & McKay, 1981, 1982; Gasson, 1983). The molecular basis of this event is presently being intensively studied both here and in McKay's laboratory in the USA. It is further described in Section 4. From an evolutionary standpoint it is of interest that cell aggregation is also an established feature of a sophisticated inducible conjugation system found to operate in some *Str. faecalis* strains (Clewell, 1981a). It will be interesting to determine to what extent conjugation associated with those plasmids from medical isolates is related to lactose plasmid conjugation in dairy starter strains.

A quite different conjugation system has been introduced into lactic streptococci from just such a hospital isolate. The erythromycin resistance plasmid pAMβ1, initially characterized in *Str. faecalis* DS-5 (Clewell *et al.*, 1974) is self-transmissible to a wide variety of gram-positive species and genera. The plasmid was first transferred from a haemolysin-defective derivative of *Str. faecalis* DS-5 into *Str. lactis* 712 (Gasson & Davies, 1980a). Subsequently the plasmid has been transferred from *Str. lactis* 712 into and between other strains of *Str. lactis*, *Str. cremoris*, *Str. lactis* subsp. *diacetylactis* and *Str. thermophilus* (Gasson & Davies, 1980a; Gasson, unpublished). In addition, the plasmid has been introduced into *Lb. salivarius*, *Lb. acidophilus*, *Lb. casei* and *Lb. reuteri* (Gibson *et al.*, 1979; Vescovo *et al.*, 1983) as well as *Bacillus subtilis* (Clewell, 1981b) and *Staphylococcus aureus* (Engel *et al.*, 1980). The wide range of host strains in which the plasmid's replication and transfer phenotypes are expressed make pAMβ1 a potentially very useful plasmid both for *in vivo* strain development and as the basis of vector construction. The efficiency of pAMβ1 transfer is however very variable. In some cases transfer frequencies are extremely low, as in many *Str. thermophilus* strains (Gasson & Davies, 1980b; Gasson, unpublished) or transfer may not be detected at all as has been found for *Lb. bulgaricus* (Vescovo, Morelli & Gasson, unpublished) and *Pediococcus* (Gonzalez & Kunka, 1983). One approach to the problem is to isolate increased recipient ability mutants, and this proved successful in raising pAMβ1 transfer into *Str. thermophilus* 1526 from $4 \cdot 2 \times 10^{-7}$ to $2 \cdot 9 \times 10^{-2}$ progeny per recipient (Gasson, unpublished). As an alternative, a different wide host range transmissible drug-resistance plasmid has been tried. Gonzalez & Kunka (1983) were able to transfer the chloramphenicol and erythromycin resistance plasmid pIP501 into and between strains of *Pediococcus pentosaceus* and *Pediococcus acidilaciti* even though pAMβ1 transfer did not occur.

Various bacteriocin plasmids and the plasmid-like gene blocks for sucrose utilization and nisin production and for diplococcin production are able to effect their own transfer by conjugation (Davey & Pearce, 1982; McKay & Baldwin, 1982; Scherwitz *et al.*, 1983; Neve *et al.*, 1984; Gasson, 1984a).

Conjugation has considerable potential for the construction of new dairy starter strains. There is no theoretical limit to the amount of DNA that can be transferred and the promiscuous nature of pAMβ1 conjugation makes transfer between widely different strains a possibility. However, it is not yet clear to what extent the various conjugation systems can be used to mobilize other plasmids and the bacterial chromosome. The lactose plasmids can certainly mobilize non-self transmissible cryptic plasmids in *Str. lactis* but similar information for the potentially more useful wide host range plasmids is not available. Finally, a novel type of transposon which is also capable of promoting conjugal transfer has been discovered and characterized in *Str. faecalis* strains (Franke & Clewell, 1981).

The transmissible tetracycline resistance transposon Tn916 has been transferred into *Str. lactis* M1 by L. Pearce (personal communication) and there is considerable potential for exploiting this novel type of gene transfer system, which combines transfer with transposition, in the lactic streptococci.

3.3. Transformation and Transfection

The introduction of purified DNA into viable bacteria is a particularly important gene transfer process because it forms a vital link in the gene-cloning technique. Attempts to achieve transformation by establishing competence in whole cells have not been successful (C. Daly, personal communication) and research is now mainly directed towards DNA uptake by bacterial protoplasts. Protoplast production and regeneration involves cell wall digestion with lysozyme (Gasson, 1980), mutanolysin (Kondo & McKay, 1982a) or a combination of amylase and lysozyme (Okamato *et al.,* 1983). So far, transformation has only been achieved by Kondo & McKay (1982b) who used polyethylene glycol treatment of *Str. lactis* ML3 protoplasts to introduce a transductionally shortened lactose plasmid. Although this initial report was of very inefficient transformation, considerably improved results are now being obtained with frequencies of about 10^5 transformants per μg of DNA (L McKay, personal communication). Polyethylene glycol treated protoplasts have also been used to effect the transfection of bacteriophage DNA into *Str. lactis* subsp. *diacetylactis* F712 (Geis, 1982).

3.4. Protoplast Fusion

Gene transfer has also been achieved by using polyethylene glycol to fuse protoplasts together. Transfer of both plasmid and chromosomal genes between genetically marked derivatives of *Str. lactis* 712 has been reported (Gasson, 1980) and some success with more difficult *Str. cremoris* starter strains has been achieved by French researchers (R. Osborne, personal communication). In prokaryotes protoplast fusion is a unique gene transfer process because it is non-specific and involves total genome contact rather than unidirectional transfer of pieces of DNA. For this reason it should be a very useful technique for the development of new starter strains. Properties of interest for starter usage may often be very difficult to define in precise genetic terms and may be controlled by multiple genes, possibly spread around the chromosome. The crude genetic mixing possible by protoplast fusion may ultimately allow such complex traits to be transferred without ever needing to understand them in a scientific way.

3.5. Gene Cloning Technology

The gene cloning technique is at the opposite end of the scale in terms of genetic sophistication. It provides the opportunity of isolating individual genes from starter bacteria or from a completely unrelated organism, and re-introducing them in a precise controlled way to create new starter strains. Genetic engineering has not yet been used in lactic streptococci or lactobacilli relevant to the food and dairy industries. The various essential components of the process are, however, becoming available. Efficient *Str. lactis* transformation has been reported by one group (L. L. McKay, personal communication) and is being sought in various other laboratories. Kok *et al.* (1983) have already developed a useful vector by cloning chloramphenicol and erythromycin resistance genes into a small cryptic plasmid from *Str. cremoris* Wg2. The resultant hybrid molecule has unique sites for restriction endonucleases *Bcl* I and *Bgl* II allowing the use of insertional inactivation. Interestingly, the plasmid can replicate both in *Bacillus subtilis* and *Escherichia coli* making it valuable as a multiple host shuttle vector. Ultimately such classical cloning vectors, using drug-resistance genes, may be considered unethical or unsafe for direct use in food manufacture so that vectors constructed entirely from metabolic plasmids in lactic streptococci or lactobacilli will also be required. Molecular analysis of suitable metabolic plasmids and isolation of relevant starter genes by cloning into other species with already established systems for genetic engineering is underway in a number of laboratories (see Section 4).

4. MOLECULAR ANALYSIS OF PLASMIDS AND GENE TRANSFER

Restriction endonuclease mapping is a powerful analytical technique for determining the relationships between, and the changes occurring in, plasmid molecules. Combined with deletion mapping and cloning, it allows specific genetic functions to be located on DNA and has also facilitated molecular study of some gene transfer processes.

4.1. Restriction Endonuclease Mapping

Restriction endonuclease mapping data are now available for a variety of plasmids which have known metabolic functions and for a bacteriophage capable of transfection.

The map for pCT176, the citrate plasmid of *Str. lactis* subsp. *diacetylactis* 176, is illustrated in Fig. 4. This plasmid, which seems identical to citrate plasmids from five other strains (Kempler &

McKay, 1981; Gasson, 1984b) could be used as the basis of a metabolic vector. It is the only small multicopy plasmid with a known function, its phenotype is readily detected on a simple indicator medium (Kneteman, 1952; Kempler & McKay, 1980) and a plasmid-free derivative of its host strain is available (Gasson, 1984b). The plasmid restriction map reveals unique or closely paired sites for restriction endonucleases *Bam* HI, *Bcl* I, *Bgl* II, *Cla* I, *Eco* RI and *Xho* I all of which are suitable for gene cloning (Gasson, 1984a, see Table 1).

The genome of bacteriophage P008 consists of a 19·6 Mdal double stranded DNA molecule with cohesive ends. It has been restriction mapped by Loof *et al.* (1983) and unique sites were found for restriction endonucleases *Bcl* I, *Bgl* II and *Hind* III. Because this bacteriophage DNA can be transfected into protoplasts (Geis, 1982), P008 may have potential for use as a bacteriophage vector.

The most intensive molecular characterization involves the lactose and proteinase plasmid pLP712 from *Str. lactis* 712 (Gasson *et al.*, 1982; Gasson, 1983; Gasson & Warner, 1982) and the related molecules in *Str. lactis* C2 and ML3 (Walsh & McKay, 1982; Gasson, unpublished; McKay, personal communication). The 55 kb plasmid pLP712 from *Str. lactis* 712 has now been analysed with 38 different restriction endonucleases and a comprehensive map has been constructed (Gasson, 1983 and unpublished). Deletions generated from pLP712 have been positioned on this map and, by taking into account the phenotype of the shortened plasmids, regions of the map which control lactose utilization, proteinase production and plasmid replication have been defined (Gasson, 1983 and unpublished).

Kok *et al.* (1983) have restriction mapped the 17 Mdal proteinase plasmid pWV05 from *Str. cremoris* Wg2 using 7 different restriction endonucleases. Southern hybridization against a different proteinase plasmid from another strain was used to define a likely location for the proteinase gene(s) on pWV05 (Kok *et al.*, 1983; Kok & Venema, personal communication).

4.2. Cloning Genes from Lactic-acid Bacteria

Although genetic engineering in lactic streptococci and lactobacilli has yet to begin, the cloning, and in some cases the expression of genes from starter bacteria in alternative hosts has been achieved.

The streptococcal vector pDB101 was used by Harlander & McKay (1982) to clone lactose genes from *Str. lactis* C2 into lactose-negative mutants of *Str. sanguis* Challis. The cloned *Str. lactis* genes complemented both a phospho-β-galactosidase mutant and a phosphotransferase system mutant of the new host. Lee *et al.* (1982) constructed an *E. coli* gene library from the *Lb. casei* 64H lactose plasmid and were able to isolate a clone in which the *Lactobacillus* phospho-β-galactosidase gene was expressed. Enzyme activity was detected and *E. coli* minicells were used to demonstrate the presence of a relevant sized radiolabelled protein. The latter technique also revealed the presence of a second expressed *Lactobacillus* gene for which a function is not presently known. As well as seeking *Lactobacillus* genes specifying components of the phosphotransferase system, Chassy's group are also hoping to sequence the *Lactobacillus* phospho-β-galactosidase gene and thus reveal the *Lactobacillus* gene expression signals (B. Chassy, personal communication). The lactose genetic region from the *Str. lactis* plasmid pLP712 has also been cloned into *E. coli*. This was achieved by isolating a fragment known from the restriction and deletion map of the plasmid to encode lactose genes. Subsequently the streptococcal phospho-β-galactosidase enzyme activity was detected and used to precisely locate this gene (Maeda & Gasson, unpublished).

Attempts have also been made to clone the region of pLP712 that controls proteinase activity, again by isolating a fragment defined on the deletion map. Use of the multicopy *E. coli* vector pAT153 has thus far proved unsuccessful (Gasson, unpublished). Kok & Venema (personal communication) have had a similar experience in attempting to clone the proteinase gene(s) from the *Str. cremoris* plasmid pWV05 into *E. coli*. It is possible that expression of proteinase genes in *E. coli* is lethal, and both research groups are now using alternative cloning strategies to overcome the problem.

The citrate plasmid pCT176 has been cloned in its entirety and as various segments into *E. coli* (Gasson, 1984b) and expression systems available in this host are being used to locate genes on the citrate plasmid restriction map.

4.3. Molecular Analysis of Transduction

Transduction of lactose genes in *Str. lactis* C2, 712 or ML3 creates novel lactose plasmids which generally match the size of the bacteriophage genome. These plasmids are transduced at elevated frequencies in subsequent transfer experiments and may or may not also carry genes for proteinase production (McKay & Baldwin, 1974; McKay *et al.*, 1973, 1976; Davies & Gasson, 1981). For *Str. lactis* 712 restriction endonuclease analysis has been used to compare the bacteriophage

DNA, the pLP712 plasmid DNA and numerous Lac^+Prt^+ and Lac^+Prt^- transduced plasmids (Gasson, 1982, 1983; Gasson & Warner, 1982 and unpublished). Transduced plasmids appeared to be deletions of pLP712 and carried no bacteriophage DNA. Lac^+Prt^+ transductants generated in a plasmid-free background were identical and carried a single deletion. In contrast, Lac^+Prt^- transductants varied and carried two or more deletions. These observations can be interpreted to provide a simple explanation for the transduction process. The lactose plasmid, pLP712, is inherently unstable, readily undergoing spontaneous deletion (Gasson, 1982, 1983). After induction of a lysogen, a spontaneously deleted lactose plasmid of a suitable size may mistakenly be packaged into a bacteriophage head. Transduction experiments would then select the deleted lactose plasmid. In subsequent rounds of transduction, erroneous packaging of the already deleted plasmid would be more frequent, causing increased transduction frequencies and accounting for the HFT phenomenon. The restriction and deletion map of pLP712 may explain the multiple deletions characteristic of Lac^+Prt^- transductants. Single deletions are unlikely because of the position of the proteinase genes between the lactose genes and the plasmid replication region. If such a deletion was big enough to facilitate packaging into a phage head, then either the lactose genes or the plasmid replicator may also be damaged, thus eliminating lactose plasmid transduction.

A variety of more unusual molecular events have been noted after lactose plasmid transduction. As well as insertion into the chromosome (McKay & Baldwin, 1978; Davies & Gasson, 1981) large novel lactose plasmids of 44 Mdal and 55 Mdal have been observed after transduction of lactose genes in *Str. lactis* 712 strains with complete cryptic plasmid complements (Davies & Gasson, 1981). Molecular analysis of the lactose plasmids after their transfer into a plasmid-free background suggests that they were created after transduction by recombination with a cryptic plasmid present in the recipient strain (Gasson *et al.*, 1982; Warner & Gasson, unpublished).

4.4. Molecular Analysis of Lactose Plasmid Conjugation

Conjugal transfer of lactose plasmids from *Str. lactis* ML3 to a plasmid-free *Str. lactis* C2 strain and between plasmid-free *Str. lactis* 712 derivatives can generate variant strains with very high donor ability and constitutive expression of a cell aggregation phenotype. The application of alkaline denaturation lysis methods to these variant strains has revealed the presence of enlarged lactose plasmids (Walsh & McKay, 1981, 1982; Gasson, 1983; Maeda & Gasson, unpublished). Enlarged plasmids are also sometimes found in progeny strains which have not undergone a change of phenotype (Walsh & McKay, 1982; Maeda & Gasson, unpublished). An example of each type of enlarged plasmid was compared with the natural lactose plasmid of *Str. lactis* ML3 by Walsh & McKay (1982). The enlarged molecules consisted of the whole lactose plasmid with different insertions of completely novel DNA.

Recently, two detailed analyses of these events have been undertaken and whilst a complete model for lactose plasmid conjugation has yet to be put forward, some very interesting results have been found. Anderson & McKay (1983b, 1984a) used recombination deficient *Str. lactis* ML3 strains to show the insertion events to be independent of the host cell's generalized recombination system. Restriction endonuclease mapping of a large number of enlarged lactose plasmids both with and without the variant phenotype led to a conclusion about the molecular events involved. Inserted DNA varied but nonetheless, also appeared to be related. Anderson & McKay (1984a) concluded that the novel DNA was originally in the form of a circular molecule which could fuse at various points on itself with a fixed position on the lactose plasmid. The fusion position on the inserted circular molecule dictated whether the resultant hybrid plasmid would aggregate and/or transfer at high frequency. This elegant study suggested the existence of a plasmid-encoded transfer factor operating in a quite unique way to generate enlarged plasmids with varied phenotypes.

An independent investigation of the related phenomenon in *Str. lactis* 712 has also been undertaken. Maeda and Gasson (unpublished) used a 23·7 kb mini-lactose plasmid, created by introducing two deletions into pLP712, to provide a small molecule more suited to the isolation and analysis of novel inserted DNA. Restriction endonuclease analysis of a large number of enlarged progeny plasmids revealed that the insert always occurred at a fixed point within a 3 kb region of the pLP712 restriction map. As was found by Anderson & McKay (1984a), the inserted DNA varied but appeared to be related. Whilst the mapping of inserted DNA into a circular molecule has yet to be achieved in *Str. lactis* 712, one additional significant observation was made. The size of the inserted DNA varied and amongst the different inserts family groups could be defined in which the inserted DNA had one common end and one varying end. This varying end accounted for the different sizes of the inserts. One urgent question currently being studied is where the novel inserted DNA comes from. It may be a plasmid which is difficult to isolate (Anderson & McKay, 1984a), a transposable transfer factor located in the bacterial chromosome (Walsh & McKay, 1982) or even both of these. Maeda & Gasson (unpublished) have cloned *Pst*-I fragments from the inserted DNA of an enlarged progeny plasmid into the ampicillin gene of the *E. coli* vector pAT153. These cloned fragments provide an ideal probe which is being used to search for the homologous insert DNA in a donor strain before conjugation has occurred. As yet these fascinating but complex observations have not been fully explained. Clearly, they will eventually reveal the nature of the most complex bacterial conjugation system so far discovered.

4.5. *In vivo* Rearrangement of DNA in Lactic Streptococci

Whilst the DNA of a given lactic streptococcal strain consists of a bacterial chromosome and a complement of plasmids, this arrangement is not static. Changes can occur by *in vivo* processes. The lactose plasmid of *Str. lactis* 712 deletes very readily. Deletions have been isolated as spontaneous events, after growth with acriflavine or at elevated temperatures, after protoplast regeneration, after repeated sub-culture in milk, after conjugal transfer and of course they are selected by transduction. Lactose and proteinase genes from the lactose plasmid of *Str. lactis* 712, C2 and *Str. cremoris* C3 can become inserted into the chromosome after transduction. Enlarged lactose plasmids are sometimes created by insertion of novel DNA during conjugation and enlarged plasmids have been found after transduction and repeated sub-culture in milk. Lactose genes have recently been shown to move from an enlarged conjugally-generated lactose plasmid onto a 8·5 kb cryptic plasmid in a *Str. lactis* ML3 derivative lacking a generalized recombination system (Anderson & McKay, 1984b). In some of these changes very precise molecular events are occurring. Enlarged lactose plasmids generated by conjugation are created by DNA insertion at one point on the lactose plasmid. All Lac^+Prt^+ transduced plasmids isolated in plasmid-free *Str. lactis* 712 strains had identical deletions. Ninety-five precent of Lac^-Prt^+ deletions generated by protoplast regeneration appeared to be identical.

These ordered molecular interactions strongly suggest that insertion sequences and transposons may be present in the lactic streptococcal genome. In this regard it is interesting to note that hybridization experiments have recently revealed weak homology between apparently different plasmids. Bacteriocin plasmids were found by Neve *et al.* (1984) to weakly hybridize with various cryptic plasmids and Kok & Venema (personal communication) made the same observation when using Southern hybridization to study proteinase plasmids.

5. APPLICATIONS FOR GENETIC MANIPULATION

Eventually the genetics of lactic-acid bacteria will be applied to improve or produce new dairy starter strains. There is considerable controversy over exactly when and to what extent predicted applications will be realized. It is our view that as genetic techniques develop they will gradually be used for applied objectives which will begin to be seen in 5 to 10 years. There are no shortcuts and it will be essential to maintain a large research effort if significant advances are ever to be achieved by genetic manipulation. Also it will be necessary for the dairy industry to accept the use of defined strain starter systems. Our current knowledge of genetics in lactic-acid bacteria allows us to speculate on what may be possible.

Two plasmid-encoded mechanisms of phage resistance, restriction modification and adsorption blocking, have been discovered. The genes for these properties will be introduced into a variety of starter strains either on their existing plasmids or after their isolation by gene cloning. A system of starter strains with much improved protection from bacteriophage attack could result.

Stabilization of lactose and protein hydrolysing enzymes by insertion into the chromosome has been achieved in *Str. lactis* C2 and 712 and this approach may be extended to generally stabilize starters and eliminate slow variants.

The proteolytic enzymes and their roles in cheese maturation and off-flavour production will be better defined. Genes for different proteinases will be exchanged between strains, and control over the levels of gene expression will be possible. The most rewarding development may be the creation of strains capable of accelerating the ripening of Cheddar cheese either by manipulating existing starter genes or by the introduction of new ones from non-dairy sources.

Starter failure due to antibiotic residues in milk could be eliminated by the isolation of drug resistant starter strains. Understanding the genetics of nisin production by *Str. lactis* may lead to the development of new strains with increased levels of antibiotic production. Protoplast fusion may allow such traits as lyophilization resistance to be transferred to new strains thus increasing the range of starters available for direct vat inoculation. Plasmid profiles will be more widely used for strain identification and for determining relationships between different starter cultures.

The application of gene cloning technology to lactic acid bacteria is the most likely process to create a major development in terms of new industrial processes. As well as being a valuable approach for starter improvement, it offers the possibility of using lactic acid bacteria to produce completely novel value added products. The most frequently quoted example is of calf rennet, the gene for which has already been cloned in *E. coli* and yeast. Production of cloned rennet by a milk fermentation process has the obvious attraction of being safe and unlikely to cause off-flavours. Other possibilities will undoubtedly arise as biotechnology progresses, and a further application area will be in the use of lactobacilli to convert whey into useful products.

6. CONCLUSIONS

Over the past ten years or so the dairy lactic-acid bacteria have progressed from a genetically obscure, unstudied group of microbes, through the pioneering work of McKay's group in the USA, to the present position with new research groups appearing in all parts of the world. The next few years will see the application of powerful molecular biological techniques and, hopefully, the use of genetic engineering in lactic-acid bacteria. This will ensure the further understanding of the many interesting genetic phenomena that have been discovered in recent years. The already available techniques of conventional genetic analysis may soon be used to produce strains of value to the dairy industry. For instance, plasmid transfer may yield starters with improved bacteriophage resistance or new proteolytic activities. Major application of genetic manipulation will need more time but almost certainly useful developments will occur. It should be remembered that whilst the potential of genetic engineering is enormous, the dairy starter bacteria have barely been investigated. Geneticists are not equipped with magical powers and the industry is not about to be revolutionized by the instant appearance of 'superbugs'. Considering the very limited research effort that starter bacteria have received,

124   *Michael J. Gasson and F. Lyndon Davies*

progress has been quite remarkable and the future of genetic research is assured. The dairy industry can now claim, rather belatedly, to have joined the world of biotechnology.

REFERENCES

ALLEN, L. K., SANDINE, W. E. & ELLIKER, P. R. (1963) *J. Dairy Res.* **30**, 351.
ANDERSON, D. G. & McKAY, L. L. (1977) *J. Bacteriol.* **129**, 367.
ANDERSON, D. G. & McKAY, L. L. (1983a) *Appl. Environ. Microbiol.* **46**, 549.
ANDERSON, D. G. & McKAY, L. L. (1983b) *J. Bacteriol.* **155**, 930.
ANDERSON, D. G. & McKAY, L. L. (1984a) *J. Bacteriol.* (In press).
ANDERSON, D. G. & McKAY, L. L. (1984b) *Appl. Environ. Microbiol.* **47**, 243.
CHASSY, B. M., GIBSON, E. M. & GIUFFRIDA, A. (1978) *Current Microbiol.* **1**, 141.
CLEWELL, D. B. (1981a) *Microbiol. Rev.* **45**, 409.
CLEWELL, D. B. (1981b) In *Molecular Biology, Pathogenicity and Ecology of Bacterial Plasmids,* R. C. Clowes, S. B. Levy & E. L. Koenig, Ed. Plenum Press, New York, p. 191.
CLEWELL, D. B., YAGI, Y., DUNNY, G. M. & SCHULTZ, S. K. (1974) *J. Bacteriol.* **117**, 283.
CORDS, B. R. & McKAY, L. L. (1974) *J. Bacteriol.* **119**, 830–9.
CORDS, B. R., McKAY, L. L. & GUERRY, P. (1974) *J. Bacteriol.* **117**, 1149.
CROW, V. L., DAVEY, G. P., PEARCE, L. E. & THOMAS, T. D. (1983) *J. Bacteriol.* **153**, 76.
DAVEY, G. P. & PEARCE, L. E. (1982) In *Microbiology 1982,* D. Schlessinger, Ed. American Society for Microbiology, Washington, DC, p. 221.
DAVIES, F. L. & GASSON, M. J. (1981) *J. Dairy Res.* **48**, 363.
DAVIES, F. L., UNDERWOOD, H. M. & GASSON, M. J. (1981) *J. Appl. Bacteriol.* **51**, 325.
DEMKO, G. M., BLANTON, S. J. B. & BENOIT, R. E. (1972) *J. Bacteriol.* **112**, 1335–45.
DOBRZANSKI, W. T., BARDOWSKI, J., KOZAK, W. & KAJDEL, J. (1982) In *Microbiology 1982,* D. Schlessinger, Ed. American Society for Microbiology, Washington, DC, p. 225.
EFSTATHIOU, Y. D. & McKAY, L. L. (1977) *J. Bacteriol.* **130**, 257.
ENGEL, H. W. B., SOEDIRMAN, N., ROST, J-, VAN LEEUWEN, W. & VAN EMBDEN, J. D. A. (1980) *J. Bacteriol.* **142**, 407.
FRANKE, A. E. & CLEWELL, D. B. (1981) *J. Bacteriol.* **145**, 494.
FUCHS, P. G., ZAJDEL, J. & DOBRZANSKI, W. T. (1975) *J. General Microbiol.* **88**, 189.
GARVIE, E. I. & MABBITT, L. A. (1956) *J. Dairy Res.* **23**, 305.
GASSON, M. J. (1980) *FEMS Microbiol. Lett.* **9**, 99.

GASSON, M. J. (1982) In *Microbiology 1982*, D. Schlessinger, Ed. American Society for Microbiology, Washington, DC, p. 217.

GASSON, M. J. (1983) *J. Bacteriol.* **154**, 1.

GASSON, M. J. (1984a) *FEMS Microbiol. Lett.* **21**, 7–10.

GASSON, M. J. (1984b) *Plasmid* (In press).

GASSON, M. J. & DAVIES, F. L. (1979) *Soc. General Microbiol. Quarterly* **6**, 87.

GASSON, M. J. & DAVIES, F. L. (1980a) *FEMS Microbiol. Lett.* **7**, 51.

GASSON, M. J. & DAVIES, F. L. (1980b) *J. Bacteriol.* **143**, 1260.

GASSON, M. J. & WARNER, P. J. (1982) *Irish J. Dairy Food Sci.* **6**, 206.

GASSON, M. J., WARNER, P. J. & ELKINS, J. (1982) *Soc. General Microbiol. Quarterly* **9**, M8.

GEIS, A. (1982) *FEMS Microbiol. Lett.* **15**, 119.

GIBSON, E. M., CHACE, S. B., LONDON, S. B. & LONDON, J. (1979) *J. Bacteriol.* **137**, 614.

GONZALEZ, C. F. & KUNKA, B. S. (1983) *Appl. Environ. Microbiol.* **46**, 81.

HARLANDER, S. K. & McKAY, L. L. (1982) *Am. Dairy Sci. Abstr.* 1982, 52.

HARRIMAN, L. A. & HAMMER, B. W. (1931) *J. Dairy Sci.* **14**, 40.

KEMPLER, G. M. & McKAY, L. L. (1979a) *Appl. Environ. Microbiol.* **37**, 316.

KEMPLER, G. M. & McKAY, L. L. (1979b) *Appl. Environ. Microbiol.* **37**, 1041.

KEMPLER, G. M. & McKAY, L. L. (1980) *Appl. Environ. Microbiol.* **39**, 926.

KEMPLER, G. M. & McKAY, L. L. (1981) *J. Dairy Sci.* **64**, 1527.

KLAENHAMMER, T. R., McKAY, L. L. & BALDWIN, K. A. (1978) *Appl. Environ. Microbiol.* **35**, 592.

KNETEMAN, A. (1952) *Antonie van Leeuwenhoek* **18**, 275.

KOK, J., DE VOS, W. M., VOSMAN, B., VAN DIJL, J. M. & VENEMA, G. (1983) Commission of the European Communities, Research and Training Programme in Biomolecular Engineering, Book of Abstracts, p. 28.

KONDO, J. K. & McKAY, L. L. (1982a) *J. Dairy Sci.* **65**, 1428.

KONDO, J. K. & McKAY, L. L. (1982b) *Appl. Environ. Microbiol.* **43**, 1213.

KOZAK, W., RAJCHERT-TRZPIL, M. & DOBRZANSKI, W. T. (1974) *J. General Microbiol.* **83**, 295.

KUHL, S. A., LARSEN, L. O. & McKAY, L. L. (1979) *Appl. Environ. Microbiol.* **37**, 1193.

LARSEN, L. D. & McKAY, L. L. (1978) *Appl. Environ. Microbiol.* **36**, 944.

LEBLANC, D. J. & LEE, L. N. (1979) *J. Bacteriol.* **140**, 1112.

LEBLANC, D. J., CROW, V. L., LEE, L. N. & GARON, C. F. (1979) *J. Bacteriol.* **137**, 878.

LEBLANC, D. J., CROW, V. L. & LEE, L. N. (1980) In *Plasmids and Transposons: Environmental Effects and Maintenance Mechanisms*, W. Stuttard & P. Rozec, Ed. Academic Press, New York, p. 31.

LEE, L. J., HANSEN, J. B., JAGUSZTYN-KRYNICKA, K. & CHASSY, B. M. (1982) *J. Bacteriol.* **152**, 1138.

LOOF, M., LEMBKE, J. & TEUBER, M. (1983) *Systematic Appl. Microbiol.* **4**, 413.

McKAY, L. L. (1983) *Antonie van Leeuwenhoek* **49**, 259.

McKAY, L. L. & BALDWIN, K. A. (1974) *Appl. Microbiol.* **28**, 342.

McKay, L. L. & Baldwin, K. A. (1975) *Appl. Microbiol.* **29**, 546.

McKay, L. L. & Baldwin, K. A. (1978) *Appl. Environ. Microbiol.* **36**, 360.

McKay, L. L. & Baldwin, K. A. (1982) In *Microbiology 1982*, D. Schlessinger, Ed. American Society for Microbiology, Washington, DC, p. 210.

McKay, L. L., Miller III, A., Sandine, W. E. & Elliker, P. R. (1970) *J. Bacteriol.* **102**, 804–9.

McKay, L. L., Baldwin, K. A. & Zottola, E. A. (1972) *Appl. Microbiol.* **23**, 1090.

McKay, L. L., Cords, B. R. & Baldwin, K. A. (1973) *J. Bacteriol.* **115**, 810.

McKay, L. L., Baldwin, K. A. & Efstathiou, J. D. (1976) *Appl. Environ. Microbiol.* **32**, 45.

McKay, L. L., Baldwin, K. A. & Walsh, P. M. (1980) *Appl. Environ. Microbiol.* **40**, 84.

Meyers, J. A., Sanchez, D., Ewell, L. P. & Falkow, S. (1976) *J. Bacteriol.* **127**, 1529–37.

Neve, H., Geis, A. & Teuber, M. (1984) *J. Bacteriol.* **157**, 833.

Okamato, T., Fujita, Y. & Irie, R. (1983) *Agric. Biologic. Chem.* **47**, 259.

Otto, R., de Vos, W. M. & Gavrieli, J. (1982) *Appl. Environ. Microbiol.* **43**, 1272.

Park, Y. H. & McKay, L. L. (1982) *J. Bacteriol.* **149**, 420–5.

Pearce, L. E. (1970) *Proc XVIII Int Dairy Congress*, Sydney, Vol. 1E, p. 118.

St Martin, E. J., Lee, L. N. & LeBlanc, D. J. (1982) In *Microbiology 1982*, D. Schlessinger, Ed. American Society for Microbiology, Washington, DC, pp. 232–3.

Sandine, W. E., Elliker, P. R., Allen, L. K. & Brown, W. C. (1962) *J. Dairy Sci.* **45**, 1266.

Scherwitz, K. M., Baldwin, K. A. & McKay, L. L. (1983) *Appl. Environ. Microbiol.* **45**, 1506.

Sherman, J. M. & Hussong, R. V. (1937) *J. Dairy Sci.* **20**, 101.

Smiley, M. B. & Fryder, V. (1978) *Appl. Environ. Microbiol.* **35**, 777.

Snook, R. J. & McKay, L. L. (1981) *Appl. Environ. Microbiol.* **42**, 904.

Snook, R. J., McKay, L. L. & Ahlstrand, G. G. (1981) *Appl. Environ. Microbiol.* **42**, 897.

Somkuti, G. A. & Steinberg, D. H. (1981) *J. Dairy Sci.* **64**, Suppl. 1, 66.

Teuber, M. & Lembke, J. (1983) *Antonie van Leeuwenhoek* **49**, 283.

Thompson, J. (1980) *J. Bacteriol.* **118**, 329–33.

Vescovo, M., Bottazi, V., Sarra, P. G. & Dellagio, F. (1981) *Microbiologica* **4**, 413.

Vescovo, M., Morelli, L., Bottazzi, V. & Gasson, M. J. (1983) *Appl. Environ. Microbiol.* **46**, 753.

Walsh, P. M. & McKay, L. L. (1981) *J. Bacteriol.* **146**, 937.

Walsh, P. M. & McKay, L. L. (1982) *Appl. Environ. Microbiol.* **43**, 1006.

Yu, R. S. T., Hung, T. V. & Azad, A. A. (1982) *Australian J. Dairy Technol.* **37**, 99.

Chapter 5

Bacteriophages of Dairy Lactic-acid Bacteria

F. LYNDON DAVIES and MICHAEL J. GASSON

National Institute for Research in Dairying, Shinfield, Reading, UK

1. INTRODUCTION

A wide variety of fermentation industries has experienced problems due to bacteriophage infection (Ogata, 1980). The dairy industry has recognized phages as a problem since the pioneering work of White-head in New Zealand during the 1930s. While numerous studies over the last half-century have led to some degree of control, it is unfortunately the case that phages remain a major problem in cheese-manufacture today (Walker *et al.*, 1981). Virulent phages attack starter strains in the vat, causing slow fermentation or even complete starter failure with consequent loss of the make. Measures such as the physical protection and chemical disinfection of vats and their immediate environment, the rotation of starter cultures and the judicious selection of strains have served only to contain the problem within manageable limits. Recent research in this area has been directed largely towards the differentiation of phages on the basis of ultrastructure, serology and genome characteristics, the phenomenon and implications of lysogeny among starter strains, the possibilities for developing phage resistant strains and the development of defined strain starter systems which it is claimed offer a greater measure of control over bacteriophage infection.

2. DIFFERENTIATION OF PHAGES

2.1. Ultrastructure

With few exceptions, phages of lactic-acid bacteria fall into morphology groups A or B of Bradley's classification (Bradley, 1967), having prolate or isometric heads with tails which in some cases are contractile and may have a variety of terminal fibres or plates in addition to collars and other structures along their length. Typical morphologies of these phages are given in Table 1.

TABLE 1

Morphology of Typical Bacteriophages of Dairy Lactic-acid Bacteria

Host	Head	Collar	Tail	Bradley group
Str. lactis and Str. cremoris	Isometric, 45–65 nm or	±	Non-contractile, 100–250 nm	B
	Prolate, 55–65 × 40–48 nm	±	Non-contractile, 80–110 nm	B
Str. thermophilus	Isometric, 50–70 nm	−	Non-contractile, 200–300 nm	B
Lb. helveticus	Isometric, 49–56 nm	−	Contractile (with sheath), 150–230 nm	A
Lb. bulgaricus and Lb. lactis	Isometric, 44–55 nm	+	Non-contractile, some with cross-bars, 120–215 nm	B
Leuc. dextranicum	Prolate, 53 × 47 nm	−	Non-contractile, 120 nm	B
Leuc. mesenteroides	Isometric, 55 nm	−	Non-contractile, 135 nm	B

+, collar present; −, collar absent; ±, presence of collar variable.

2.1.1. Phages of Mesophilic Streptococci

Almost all *Str. lactis* and *Str. cremoris* phages so far reported conform to Bradley's group B. Typically they have isometric heads of 45–65 nm diameter with tails 100–250 nm long or, less frequently, prolate heads

(55–65 nm × 40–48 nm) with shorter tails (80–110 nm long) (Keogh & Shimmin, 1974; Terzaghi, 1976; Tsaneva, 1976; Sozzi, 1977; Lawrence, 1978; Heap & Jarvis, 1980; Lembke *et al.* 1980; Sozzi *et al.*, 1980). An unusual phenomenon observed by Bradley (1967) was the existence of 2 distinct head lengths in preparations of a prolate headed phage of *Str. lactis* ML3. Subsequently, Keogh & Shimmin (1974) also reported variable head lengths, though not well defined dimorphism, for a phage of *Str. lactis* C2 (these 2 strains of *Str. lactis*—C2 and ML3—are now known to have common ancestry, see Davies *et al.*, 1981). Phages with exceptionally long tubular heads were recently observed by Chopin & Rousseau (1983) to occur among prolate phages of 2 *Str. lactis* strains at frequencies of 2·5 and 16%. Head lengths were up to 14 times those of the normal prolate phages.

Occasionally, phages with isometric heads of 80–100 nm diameter and tails up to 500 nm in length have also been isolated (Tsaneva, 1976; Lawrence *et al.*, 1978; Lembke *et al.*, 1980). It was suggested by Lawrence *et al.* (1978) that these large isometric and also prolate phages may be mutants of small isometric phages since the latter are most common and are the type generally isolated from lysogens. This view has been challenged by Lembke *et al.* (1980), however, on the grounds of the ultrastructural diversity now becoming evident for this group of phages.

Many of the isometric and a few of the prolate headed phages possess distinct collars and also baseplates which are in some cases very complex in structure (Keogh & Shimmin, 1974; Lawrence *et al.*, 1978; Heap & Jarvis, 1980; Lembke *et al.*, 1980). A *Str. lactis* subsp. *diacetylactis* phage (P008) commonly found in German cheese factories, was reported by Loof *et al.* (1983) to have delicate whiskers about 40 nm long emanating from its collar. Each whisker terminated in a ball-like structure 4–5 nm in diameter.

Fine structures such as these whiskers, terminal fibres and some base-plate components are close to the limits of resolution for current ultrastructural techniques and can only be differentiated in high quality negatively stained preparations. Staining conditions may need to be varied to visualize different structures (Loof *et al.*, 1983). Because different techniques may have been used and their quality is so variable, electron micrographs published by different groups frequently cannot be usefully compared. Teuber's group (Lembke *et al.*, 1980; Teuber & Lembke, 1983) have themselves, therefore, examined almost 300 isolates of phages attacking 57 strains of mesophilic streptococci in cheese factories over several years. From this extensive survey they recognized a range of 14 distinct morphological types which so far as they are able to determine would also include those published elsewhere in the literature.

Two notable exceptions to the phages usually reported for mesophilic lactic streptococci are a phage of *Str. lactis* ML1 reported by Lembke *et al.* (1980) which has a prolate head (65 × 44 nm) but a very short tail (24 nm long) with a collar extending into 4 or more fibres and a phage (KSY1) of *Str. cremoris* which has a bullet shaped head (230 nm × 50 nm), a short tail (35 nm long) and 3 collars with multiple radial projections (Saxelin *et al.*, 1979). Both of these phages have been assigned to group C of Bradley's classification. The phage KSY1 is unique in bringing about the dissolution of capsules of *Str. cremoris* (in the Finnish fermented product 'villi') even though its host strain is non-encapsulated.

2.1.2. Phages of Thermophilic Streptococci

Bacteriophages of *Str. thermophilus* strains fall into Bradley's group B and with few exceptions appear to be very similar with isometric heads (generally 50–70 mm diameter), tails 200–300 mm long, no obvious collars and only small base plates, often with a central fibre (Bauer *et al.*, 1970; Gelin *et al.*, 1970; Sozzi & Maret, 1975; Sozzi, 1977; Accolas & Spillmann, 1979a; Reinbold *et al.*, 1982). Variants with larger heads (Deane *et al.*, 1953), shorter tails (Accolas & Spillmann, 1979a) and exceptionally long tails (polytails) (Reinbold *et al.*, 1982) are also known to occur.

2.1.3. Phages of Lactobacilli

There have been relatively few ultrastructural studies of phages of lactobacilli used in dairy fermentations; even so, some morphological diversity has become apparent. Several strains of *Lactobacillus helveticus* have been examined (Sozzi & Maret, 1975; Sozzi, 1977; Accolas & Spillmann, 1979b; Reinbold *et al.*, 1982) and all show isometric heads (49–56 nm diameter) and contractile tails (150–230 nm long) with sheaths, typical of group A of Bradley's classification.

Although one strain of *Lb. bulgaricus* has recently been reported to belong to Bradley's group A (Reinbold *et al.*, 1982) all other strains of this species, and of the related *Lb. lactis* examined show Bradley group B morphology (Dentan *et al.*, 1970; Sozzi, 1977; Peake & Stanley, 1978; Accolas & Spillmann, 1979b). All reported *Lb. lactis* and *Lb.*

bulgaricus phages have isometric heads (44–55 nm diameter), tails 120–215 nm in length and collars and base plates sometimes present. Additionally, the tails of some phages of these two species bear along their length unusual cross bars (about 20 nm long and varying up to 10 in number) whose function is unknown and which have very rarely been observed on phages of any other bacterial genera (Dentan *et al.*, 1970; Sozzi, 1977; Peake & Stanley, 1978; von Kurman, 1979).

2.1.4. *Phages of Leuconostocs*

Phages of leuconostocs were not reported until relatively recently, perhaps because these bacteria tend to be slower growing and so less prone to phage attack than streptococci or because they are commonly coated with a dextran which may protect against phage adsorption (Sozzi *et al.*, 1978). A range of virulent and temperate phages against leuconostocs examined by Shin & Sato (1979b) were of two morphological types, conforming to groups A and B of Bradley (1967). Other reports of phages against *Lb. cremoris*, *Lb. dextranicum* and *Lb. mesenteroides* (Sozzi *et al.*, 1978; Shin & Sato, 1979a; Shin & Sato, 1981) have assigned them all to Bradley's morphology group B.

2.1.5. *Temperate Phages*

Since temperate phages released from lysogens may show virulence against other starter strains or mutate to virulence against their own lysogenic host (Shimizu-Kadota *et al.*, 1983), meaningful distinction between temperate and virulent phages is not always possible. Thus, of the 20 morphological types of temperate phages identified by Teuber & Lembke (1982) in lysates of mesophilic streptococci, at least 5 corresponded to those previously described as virulent phage types by Lembke *et al.* (1980). One morphological type of temperate phage occurred with particularly high frequency, being induced from each of 12 strains of lactic streptococci (Teuber & Lembke, 1983); it also appeared identical to that induced from *Str. lactis* 712 by Gasson & Davies (1980).

Until recently, all temperate phages induced from lactic streptococci were reported to be of the isometric type (e.g. Lawrence, 1978; Terzaghi & Sandine, 1981). There is now evidence, however, that although isometric forms predominate, prolate headed temperate phages may occasionally occur. For instance, after inducing a range of lactic streptococci with UV light, Heap & Jarvis (1980) detected isometric phages in lysates of 61 but prolate headed phages in only 3.

One strain of *Str. lactis* from cheese-whey was shown by Jarvis (1982) to be a multiple lysogen for 3 morphologically distinct temperate phages, one large isometric, one small isometric and one prolate-headed. An interesting observation made by Chopin *et al.* (1983) was that the ratio of isometric to prolate-headed temperate phages in lysates of a bi-lysogenic *Str. lactis* depended upon the inducing agent used.

2.2. Serology

Only the lactic streptococcal phages have been examined serologically. Jarvis (1977) subdivided phages into 6 morphological types and was able to demonstrate that phages which are similar morphologically are related serologically. With few exceptions, phages were neutralized only by antisera prepared against phages of the same morphological type. However, rates of neutralization differed so that the 6 types could be subdivided giving groups which correlated well with host-range. Subsequently, Jarvis (1978) has isolated a host range mutant of *Str. cremoris* which showed an altered rate of neutralization by an anti-phage serum.

Collars were particularly useful for differentiation since collared phages reacted most often and at higher neutralization rates with antisera prepared against collared phage and vice versa (Jarvis, 1977). No such serological specificity could be correlated with presence or absence of base plates (Jarvis, 1977). Heap & Jarvis (1980) reported that isometric and prolate phages are serologically distinct.

A different approach to serotyping of lactic phages was reported by Lembke & Teuber (1981) who used immunoelectronmicroscopy. Morphologically similar prolate phages could be divided into 5 sub-groups according to the reaction of an antiserum with head or tail components. These same authors have also described a method of detecting specific phage serotypes in whey, using an enzyme-linked immunosorbent assay (ELISA) (Lembke & Teuber, 1979).

2.3. Lytic Spectra

Only the phages of lactic streptococci have been isolated with sufficient frequency to permit schemes to be drawn up using host-range data. Such data have often been used primarily to type the host bacteria and are important in determining culture rotations. Groupings of phages based on host-range have been reported (e.g. Henning *et al.*, 1968; Nyiendo, 1974; Chopin *et al.*, 1976; Terzaghi, 1976; Boissonnet *et al.*, 1981) but are difficult to compare since different phage collections have been used. Presumably phenomena such as host controlled modification-restriction (e.g. Boussemaer *et al.*, 1980) and host-range mutation (e.g. Jarvis, 1978) affect the stability of such schemes. Sinha (1980) reported that a mutant of *Str. lactis* ML1 simultaneously showed acquisition of several new phage resistances and sensitivities.

The temperate phages from a wide range of lactic streptococci were examined by Reyrolle *et al.* (1982) and their lytic spectra found to correlate closely with those of virulent phages attacking the same strains. This corroborates the remarks made earlier concerning lack of morphological distinction between temperate and virulent phages.

While Terzaghi (1976) observed some relationship between host-range groupings and morphology, most authors have been unable to find such a correlation (Nyiendo, 1974; Tsaneva, 1976; Jarvis, 1977; Lembke *et al.*, 1980). It does, however, appear that in general prolate phages have wider host ranges than isometric phages (Terzaghi, 1976; Heap & Jarvis, 1980). For instance, 26 of 30 prolate phages but none of 20 isometric phages examined by Heap & Jarvis (1980) attacked more than 10 bacterial strains. These authors proposed that the apparent difference in host range might be explained by the fact that most temperate phages are isometric and that lysogeny confers immunity to similar phages.

2.4. Phage Genome

Twenty five phages of mesophilic lactic streptococci examined by Nyiendo (1974) all contained double-stranded DNA with guanine–cytosine contents between 32·7 and 40%; there was no obvious correlation between morphological groups and guanine–cytosine contents though some correspondence was observed between morphology and groups differentiated by DNA–DNA hybridization. More recently, Lembke & Teuber (1982) found particular types of prolate and isometric phages commonly present in cheese-wheys to have guanine–cytosine contents of 41 and 32 respectively.

By measuring the length of DNA molecules in electron-micrographs or totalling the fragment sizes of DNA digested with restriction endonucleases, genome sizes have been determined for a number of phages; those of lactic streptococci fall within the range 18–27 megadaltons (Mdal) (Klaenhammer & McKay, 1976; Daly & Fitzgerald, 1982; Loof *et al.*, 1983) and that of a lactobacillus phage was calculated as 27 Mdal (Shimizu-Kadota *et al.*, 1983).

The cleavage patterns obtained by electrophoresis of endonuclease digested DNA are highly specific and offer a sensitive approach to the identification and grouping of phages. Davies·*et al.* (1981) used this method to compare phages of *Str. lactis* and found those from the closely related strains 712 and C2 to have identical fragmentation patterns. That a virulent phage attacking the *Lb. casei* starter used in yakult manufacture was derived from a temperate phage which normally lysogenized this strain was again established very largely by the similarity of the phage genome fragmentation patterns (Shimizu-Kadota *et al.*, 1983). The technique was reported to be of high potential taxonomic value by Dhillon *et al.* (1982) in an analysis of naturally occurring temperate coliphages from diverse habitats. These authors argued that while phages cultured in the genetically isolated environments of laboratories are highly stable with regard to the number and location of enzyme sensitive base sequences, the DNA of natural populations shows variability through 'interbreeding' of phage types. The great overall variability of restriction enzyme patterns is such that isolates giving identical fragmentation patterns must be very closely and possibly clonally related. While the potential for such variations is probably more limited in the dairy environment, it is nevertheless remarkable that when Loof *et al.* (1983) examined the DNA of 10 ultrastructurally identical phages, isolated from cheese-factories over a 10 year period, they gave identical fragmentation patterns on digestion with the enzyme *Eco* R1. Thus the genetic identity of these phages had been conserved over a long period and showed good correlation with morphology.

Future attempts to probe the relationships and identities of phages will undoubtedly exploit the rapidly developing technologies for genome analysis. A good example of such a trend is the elegant study of the prophage origin of a *Lb. casei* virulent phage by Shimizu-Kadota and her co-workers at the Takult Institute in Tokyo (see Section 3.3 of this chapter). Also, Loof *et al.* (1983) have indicated that a comprehensive examination of phage relationships using restriction endonuclease analysis is underway in their laboratory. The complexity of these techniques does, however, limit their application; they are unlikely to be found in routine use. A phenomenon which could affect restriction endonuclease analysis of phage DNA has been observed by Daly & Fitzgerald (1982) and Davies & Underwood (unpublished data). When a phage is propagated on an alternative host, i.e. one which initially restricts its development but within which modification occurs to give permissive phage development, the endonuclease fragmentation pattern of the phage DNA changes significantly. If the phage is returned to its previous (and now at first restrictive) host its DNA fragmentation reverts to its former pattern on endonuclease digestion. Each pattern is, however, stable and reproducible for a phage propagated on any one host so that the analysis is valid provided the propagating host is carefully defined.

3. LYSOGENY

3.1. Incidence of Lysogens

Many strains and species of lactic-acid bacteria are lysogenic, carrying potentially inducible prophage. Following the preliminary reports by Coetzee & deKlerk (1962) on lysogeny in lactobacilli and the subsequent demonstration by deKlerk & Hugo (1970) that temperate phages could be induced from strains of *Lb. acidophilus,* many other lactobacilli (Sakurai *et al.*, 1970; Yokokura *et al.*, 1974; Stetter, 1977), lactic streptococci (Kozak *et al.*, 1973; Lowrie, 1974; Huggins & Sandine, 1979; Terzaghi & Sandine, 1981; Reyrolle *et al.*, 1982), and leuconostocs (Shin & Sato, 1980) have been shown to be lysogenic. In fact, some strains are multiple lysogens harbouring phages of two or more types (e.g. Jarvis, 1982; Chopin *et al.*, 1983); multiple lysogens are probably more common than thus far reported but are readily detectable only when the temperate phages have different morphologies.

Estimates of the frequency with which lysogens occur among the different species differ widely. For instance, Kozak *et al.* (1973) tested filtrates from cultures of *Str. cremoris* and *Str. lactis* which had been subjected to an inducing dose of ultra-violet radiation for their ability to form plaques on indicator strains and found only 8% (of 87 tested) to be lysogenic, whereas Reyrolle *et al.* (1982) using a similar approach reported that 43% of 113 such strains tested were lysogens. Davies & Gasson (1981) and Huggins & Sandine (1977) used the depression in growth curves caused by lysis of lysogenic cells following exposure to UV light to detect lysogens and reported 20% (of 69 strains) and 60% (of 63 strains) respectively to be lysogenic. The latter authors confirmed their diagnosis by demonstrating the presence of phages or incomplete phage particles in filtrates of inducible strains (i.e. presumptive lysogens) viewed under the electron microscope. Terzaghi & Sandine (1981) extended this approach, using electron microscopic examination of UV treated cultures as their primary screening method. They found intact phages, or phage heads and/or tails in each of 42 cultures examined.

We can conclude, therefore, that among the lactic streptococci most, if not all strains are lysogens and the widely ranging estimates of their frequency probably reflect: (a) the varied criteria used to determine lysogeny, (b) the subjectivity of interpreting depression of optically determined growth curves after inducing treatments, and (c) the apparent paucity of indicator strains.

3.2. Incidence of Indicator Strains

The actual frequency of indicator strains is again somewhat uncertain; it is generally thought that they are rare (e.g. Lowrie, 1974; Heap *et al.*, 1978; Terzaghi & Sandine, 1981; Davies & Gasson, 1981) though Reyrolle *et al.* (1982) found that 25% of lactic streptococci tested served as indicators. The concentration of phage lysates is clearly important in testing for indicator strains and it is also important to distinguish between inhibition due to phage and that caused by the bacteriocin like compounds produced by many lactic streptococci; dilutions of lysates should be made so as to obtain discrete plaques on indicator strains.

3.3. Practical Importance of Lysogens

While prophage may be induced from lysogens by agents such as UV light and mitomycin C (MC), it is the spontaneous liberation of temperate phages which is important in practice. These phages may show virulent activity against other strains in the same starter or strains in other starters used in the rotation. The highest levels of spontaneous phage induction (over 25% of strains tested) were recorded by Reyrolle *et al.* (1982). It is not clear to what extent UV or MC inducibility indicates the potential of a strain to liberate phage spontaneously and Meister & Ledford (1979) suggested that the cultural requirements for UV and MC induction are indeed different to those for spontaneous induction. Furthermore, it is possible that elevated temperatures, residues of detergent sterilizers (Pop & Kurmann, 1981) etc. may also have inducing properties, affecting the rate of phage liberation in commercial environments.

There have been several attempts to assess the contribution of lysogens to the level of phage infection in cheese factories. The isometric phages released from lysogenic starters are morphologically indistinguishable from those commonly isolated from cheese-wheys and in a carefully monitored trial in a New Zealand cheese factory a lytic phage attacking one starter strain was found to be similar in morphology and host-range to a phage induced by UV light from a second strain in the same rotation (Heap & Lawrence, 1977). It was also found that replacement of lysogenic strains with others not readily lysed by UV light led to more stable rotation (Lawrence *et al.*, 1978). While it seems likely from such circumstantial data that lysogenic starter strains are a major source of phages in cheese plants, there is no definitive evidence available and the difficulty in finding indicator strains is contradictory to this notion. It may be that mutation of temperate phages or operation of modification–restriction processes must occur before virulent attack of new hosts is detected.

Attempts to assess the importance of lysogens as a source of phages in cheese factories have been frustrated partly by the complexity of the starter systems in use. In a factory using a number of mixed or multiple-strain starters in rotation, there will be an environmental build-up of many strains. Recent investigations of a relatively simple system, involving phage infections of a single-strain lactobacillus starter used in the manufacture of a fermented milk drink may be seen, therefore, as a useful model to illustrate that virulent phages may be derived from prophages. *Lb. casei* strain S-1 is a lysogen harbouring temperate phage φFSW. Virulent phages (φFSVs) have been independently isolated at several factories during periods of abnormal fermentation with S-1. These φFSVs were serologically indistinguishable from each other and from φFSW (Shimizu-Kadota & Sakurai, 1982). The close relationship of the temperate and virulent phages was confirmed by a comprehensive study of their morphology, structural protein composition and DNA cleavage patterns on endonuclease digestion (Shimizu-Kadota *et al.*, 1983). Using φFSV DNA as a probe, these same authors went on to show by Southern filter hybridization that φFSV-related sequences were present in S-1 DNA and were structurally similar to those of φFSW. They therefore concluded that the virulent FSV phages were derived from the φFSW prophage in the starter strain S-1. In further genetic analysis of this system Shimizu-Kadota & Tsuchida (in press) have constructed a circular restriction map for φFSW DNA on which they have located the prophage integration site. They have also shown that the φFSW genome incorporates a host-derived sequence at a specific site, the V-region, in some of the virulent (φFSV) mutants. Since the V-region may contain the lysogeny-controlling genes of φ FSW, DNA fragments including this sequence have been cloned into a plasmid vector (pBR322) of *E. coli* for detailed analysis (Shimizu-Kadota, personal communication, Sept. 1983).

3.4. Prophage Curing of Lysogens

Super-infection immunity conferred by lysogeny is very specific and its contribution to overall phage resistance when mixed and multiple-strain systems are in use is unlikely to be significant. If, on the other hand, strains could be cured of their prophages, a potential source of infective phages would be removed. Only recently has prophage curing been reported and the ease with which it can be accomplished varies greatly between strains. Gasson & Davies (1980) described a procedure by which spontaneously cured strains could be isolated from lysogenic cultures by their ability to survive inducing doses of UV light. The procedure worked well for *Str. lactis* 712 and *Str. cremoris* R1 but was ineffective for a number of other strains of *Str. cremoris* tested. Prophage cured isolates of *Str. cremoris* R1 were also isolated on the basis of their UV non-inducibility by Georghiou *et al.* (1981) after initially irradiating the cells to effect curing. These procedures were successfully extended by Teuber & Lembke (1983) to a further strain of *Str. cremoris* from a commercial starter culture. Chopin *et al.* (1983) using the procedure of Gasson & Davies (1980) confirmed that cured derivatives were readily isolated from *Str. cremoris* R1. They found, however, that of a further 11 strains of *Str. cremoris* which did not respond to this technique, 3 were cured by simultaneous treatment with UV light and mitomycin C. *Lb. casei* S-1 cannot be induced under laboratory conditions; however, Shimizu-Kadota & Sakurai (1982) produced thermoinducible mutants from which prophage cured derivatives could readily be isolated.

There is as yet little evidence available of the benefits of using cured derivatives in commercial practice. *Lb. casei* S-1 prophage is known to mutate to virulent phages attacking the host (see Section 3.3 of this chapter). Replacement of the S-1 starter with prophage cured derivatives eliminated problems experienced in factories when the parent strain was used to produce fermented milks. Acid producing ability of the strain was unimpaired by the curing procedure (Shimizu-Kadota & Sakurai, 1982). Chopin *et al.* (1983) reported that the acid producing ability of all but one of the strains tested was unaffected by prophage curing. They also found that the ranges of virulent phages to which they were sensitive were unchanged in 3 of 4 strains after curing; the fourth strain, a bi-lysogen, became sensitive to an increased number of prolate headed virulent phages when one of its prophages was cured.

As starter systems using fewer and more precisely defined strains become more popular, there will be greater opportunity to assess the value of using prophage cured derivatives.

Although the existence of lysogens among lactic-acid bacteria has been recognized for some time, the prophage curing and subsequent relysogenization reinstating superinfection immunity described by Gasson & Davies (1980) was in fact the first formally complete demonstration of lysogeny in this group of bacteria.

4. BACTERIOPHAGE RESISTANCE

4.1. Phage Resistant Mutants

Phage resistant mutants arise spontaneously in many starter bacteria exposed to phages (e.g. Limsowtin & Terzaghi, 1976; Marshall & Berridge, 1976) and may be induced by treating cultures with mutagenic chemicals (Sinha, 1980), irradiation (Gudkov *et al.*, 1982) or both (Pyatnitsyna & Zadoyana, 1978). An approach finding favour in some countries is the propagation of starter strains in the presence of cheese-wheys from the factories in which the starters are to be used. In this way it is assumed that derivatives arise which are resistant to the phages they are likely to encounter in those factory environments (Hull, 1977; Czulak *et al.*, 1979; Huggins & Sandine, 1979; Jarvis, 1981).

Mutants derived by such procedures are, however, commonly unstable, reverting to phage sensitivity; they may also simultaneously acquire new phage sensitivities (von Frohlich *et al.*, 1978; Erickson, 1980; Sinha, 1980) or show other changes such as impaired ability to produce acid in milk. The most common mutation responsible for phage resistance is that causing an alteration of receptor sites at the cell surface (King *et al.*, 1983). This might simultaneously affect systems for transport of carbohydrates into the cell or production of surface bound proteinases, explaining the observed loss of acid producing ability. King *et al.* (1983), however, were unable to demonstrate any direct correlation between phage resistance and slow acid production in any of 10 *Str. cremoris* strains tested.

Despite potential problems of instability and altered metabolism, phage-resistant mutants can form the basis of commercial starter systems (see Section 5 of this chapter) provided that they are judiciously selected and vigilance is exercised for the early detection of new phages.

4.2. Resistance to Phage Lysins

The phenomenon of 'lysis from without' first described by Delbruck in 1940 was shown to occur among streptococci by Naylor & Czulak (1956). Phage associated lytic enzymes, lysins, can bring about lysis of phage-unrelated bacterial strains by splitting essential bonds or cross links in the peptidoglycan of their cell walls. The lysins of phage ML3 and C10 were purified by Oram & Reiter (1965) and shown to be active against many different strains of group N streptococci. The relative activities against a range of these strains was determined and the two phage lysins proved to have different specificities. Lysin from a phage infecting Str. lactis C2 was also purified by Mullan & Crawford (1981, 1982a) and shown to lyse a wide range of group N streptococci; only a few strains, notably AM2, US3 and DRC2 were relatively insensitive. These same authors (1982b) also showed that acid production by a number of phage non-related starters was inhibited when they were paired with the lysin producing strain C2, though strains AM2 and US3 were least affected. It follows that in mixed or multiple-strain starters infection of one strain by a lysin-producing phage may lead to widespread lysis, even of strains resistant to infection by that phage. There is considerable attraction, therefore, in the notion of developing lysin-resistant strains because of their potentially broad spectrum phage resistance. That such mutants can be obtained was demonstrated by Gasson (unpublished data) using lysin produced by a phage of Str. lactis ML3.

It was suggested by Verhue (1978) that the extent to which strains are naturally resistant to lysin depends upon the degree of polymerization of the cell wall peptidoglycan. He postulated that a minimum number of intramolecular bonds (either glycosidic or peptidic) has to be established for the cell to withstand the internal osmotic pressure.

Verhue (1978) also showed that low levels of penicillin can stimulate the activity of lysin due to their common cell wall target. The correlation between penicillin resistance and lysin resistance which he observed may offer an approach to selection of lysin-resistant strains.

4.3. Restriction–Modification Systems

A further type of phage resistance recognized in lactic streptococci is that determined by intracellular restriction endonucleases. Cells of a restrictive strain recognize the DNA of the invading phage as foreign and degrade it with these enzymes. However, a low level of phages survive and multiply in the cell giving rise to progeny phages with modified DNA which are fully virulent upon that strain. This phenomenon of restriction–modification (R/M) was first reported in lactic streptococci by Collins in 1956 and was more recently described in cheese starter strains by Pearce & Lowrie (1974). Boussemaer *et al.* (1980) found that of 30 non-homologous reactions between lactic streptococci and their phages, the bacteria were fully sensitive in 4 and restricted the phages in 23; using a mathematical model, 4 or 5 R/M systems were defined of which up to 3 were found in one strain. These authors concluded that R/M systems constitute one of the main defences of lactic streptococci against phages and argued that a knowledge of these is relevant to definitions of rotations of starter strains.

Limsowtin *et al.* (1978) successively transferred single strains of lactic streptococci in milk and observed phage sensitivity to be variable; in one strain this was attributed to the loss of R/M systems. The unstable nature of R/M systems subsequently led to speculation that they may be plasmid encoded and Sanders & Klaenhammer (1981) were able to demonstrate that the determinant for such a system in *Str. cremoris* KH was located on a 10 Mdal plasmid. A similar observation was made for the restrictive activity of *Str. lactis* subsp. *diacetylactis* F7/2 (Teuber & Lembke, 1983). An R/M system in *Str. lactis* 712, however, operates against a phage of *Str. cremoris* 2501 with equal efficiency in a plasmid-free derivative of the restrictive host (Davies, unpublished data) so that not all such systems are plasmid encoded.

It was suggested by Erickson (1980) that genes determining the production of different restriction endonucleases might be introduced into the genomes of starter strains, perhaps by plasmid transfer, to improve their phage resistance. A novel restriction–modification system was introduced into a plasmid free *Str. lactis* 712 strain when sucrose genes were conjugally transferred from eight different nisin-producing donor strains (Gasson, 1984) suggesting that such an approach might be feasible. Daly & Fitzgerald (1982) and Fitzgerald *et al.* (1982) described the isolation and purification of a novel sequence-specific restriction endonuclease, *Scr* F1, from *Str. cremoris* strain F.

They found that the enzyme was active against a wide range of lactic streptococcal phages and suggested that the cloning of genes for this and other restriction enzymes into selected starter strains would greatly increase their phage resistance.

Daly & Fitzgerald (1982) also examined a number of R/M systems in other *Str. cremoris* strains and showed that the endonuclease cleavage patterns of their phage DNAs were significantly altered when the phages were propagated on restrictive hosts. Their data suggested that this was due to host controlled modification of the phage DNA.

An important practical observation made by Pearce (1978) and Sanders & Klaenhammer (1980) was that restriction systems may cease to be effective at temperatures of 40°C or higher. The scald temperatures in cheesemaking are of this order so it is possible that starters could suffer a loss of restrictive activity at this stage of the process. The temperature sensitivity of restriction enzymes is probably variable, however, since that from *Str. cremoris* F1 showed optimum activity at 40°C (Daly & Fitzgerald, 1982).

4.4. Plasmid Encoded Adsorption Resistance

Two very recent investigations have demonstrated a novel type of extrachromosomally determined phage resistance in cheesemaking strains of *Str. lactis* and *Str. cremoris*. Sanders & Klaenhammer (1983, 1984) found that on serial transfer, the phage insensitive strain *Str. lactis* ME2 spontaneously gave rise to mutants sensitive to a phage, φ 18, of *Str. cremoris*. All the mutants examined had lost a 30 Mdal plasmid and adsorbed φ 18 at levels over 99·8% compared to adsorption of only 20–40% by the resistant parent strain. These authors concluded, therefore, that the 30 Mdal plasmid in ME2 encodes a mechanism which reduces the level of phage adsorption. Significantly, the loss of this plasmid also permitted more efficient adsorption of certain other lactic phages to cells of ME2, suggesting a broad spectrum phenomenon.

A second, rather similar, system has been described by de Vos *et al.* (1984) whereby a 34 Mdal plasmid in *Str. cremoris* SK11 encodes resistance to phage φ SK11G. There was plasmid heterogeneity between and within cultures of this strain but only isolates carrying the 34 MDal plasmid were resistant. Phage sensitive mutants isolated by heat and acriflavine curing were shown to have lost the plasmid and when this was re-introduced by conjugal transfer from the parent strain (de Vos, Davies & Underwood, unpublished data) phage resistance was simultaneously re-acquired. The mechanism of resistance was again shown to be reduction of phage adsorption to cells carrying the plasmid (80–90% in sensitive cells compared to 1–5% in those which were resistant). The possibility of prophage or restriction–modification involvement was ruled out.

The inhibition of phage replication by extrachromosomal genetic elements was reviewed at some length by Duckworth *et al.* (1981) but none of the mechanisms described appear to explain these recent observations for lactic streptococci. While the biochemical basis of the resistance has yet to be investigated, the use of plasmids of this type for the construction of improved starter strains is an interesting possibility.

5. PHAGE CONTROL AND STARTER CULTURE SYSTEMS

5.1. Precautionary Measures in Commercial Practice

Over the years that the bacteriophage problem has been recognized in the dairy industry, numerous measures have been adopted with varying degrees of success to prevent or contain phage development. These include methods of physical protection such as the enclosure of vats, the filtration of air within the cheese room and to starter vessels, and the physical separation of starter rooms from production areas. Vats, filling lines and ancillary equipment are sterilized by steam or chlorination and incoming milk is severely heated if to be used for bulk starter preparation. It is known that some phages of lactic-acid bacteria can survive HTST pasteurization and to a lesser extent, spray drying (Chopin, 1980). Starter cultures are generally inoculated through special seals, commonly protected by a hypochlorite barrier (e.g. Cox & Lewis, 1972). While many new disinfectants are now available, hypochlorite or other chlorine based compounds are still the most effective for phage disinfection in most circumstances.

All the above measures, subject to various modifications, have been common practice for some time and have been adequately reviewed (e.g. Wigley, 1980; Walker *et al.*, 1981; Tamine, 1981). In recent years it is through modified approaches to bulk starter preparation and the development of new starter systems that the major efforts have been made to improve phage control. These developments apply almost exclusively to the cheese manufacturing industry since phage problems in the production of yoghurt or other fermented milks are comparatively rare. The reason for this is not clear but the use of relatively few strains and the shorter history of large scale yoghurt production may be important factors.

5.2. Bulk Starter Preparations

The growth of starter cultures in milk or milk-based media is eventually limited by the lactic acid they produce and cell concentrations up to ten-fold higher can be achieved if some form of neutralization is used. This may take the form of automatic titration whereby sodium or ammonium hydroxide is added to maintain culture pH at 6·0–6·5 or buffering of the starter medium with phosphates or citrates. A whey-based medium fortified with yeast extract and buffered with phosphates was claimed to give optimal culture activity when coupled with anhydrous ammonia injection to maintain a pH of 6·0 (Richardson *et al.*, 1977). A reported advantage of buffered media is their ability to complex available calcium, suppressing the adsorption of phages to the cells. The use of calcium free media as a means of phage control was advocated by Reiter as long ago as 1956. In a survey of 7 commercially prepared 'phage-inhibitory media' Gulstrom *et al.* (1979) concluded that although the capacity of a medium to suppress phage proliferation was related to its buffering capacity, the inclusion of sufficient nutrients to ensure rapid culture growth and overcome the effect of high phosphate or citrate levels was equally important in this respect. A recently-developed whey-based and fortified bulk starter medium (Sandine & Ayres, 1981) includes an insoluble buffer system in which the components are reported to be encapsulated (Birlison & Stanley, 1982) and slowly released throughout the incubation. Thus it is claimed that the potentially damaging effects of high buffer concentrations on the starter cells (Ledford & Speck, 1979) are avoided, though calcium complexing is still achieved giving effective phage control (Willrett *et al.*, 1982).

5.3. Direct Vat Inoculation

Conventional methods of starter handling whereby master and mother cultures are prepared at the cheese factory are steadily being replaced by the use of cell concentrates purchased from a starter manufacture. These may be used for direct inoculation of the bulk starter tanks or of the cheese vat itself, eliminating some or all of the preparative stages and reducing the opportunity for phage infection. These concentrates have in the main been sold as deep-frozen cultures (Wigley, 1980) which must be maintained at low temperature throughout shipment and storage. A recent development, however, is the sale of freeze-dried starter cultures (Kringelum, 1982; Martin, 1983) which require no special conditions for shipment and storage. When starter is added directly to the cheese-vat as a freeze-dried culture there is no carry-over of acid so that some adjustment of 'make' conditions may be necessary. However, Chapman (1978) found that growth curves and the development of acidity using freeze-dried cultures were similar to those of milk grown cultures during Cheddar cheese manufacture.

5.4. Defined Strain Starter Systems
The types of starter cultures used commercially were classified by Lawrence *et al.* (1976) into 'single', 'paired' and 'multiple' strain types in which defined strains are used, and 'mixed' strain starters which are of undefined composition. While mixed strain starters have been heavily favoured in the UK, other countries, notably New Zealand and Australia, have successfully used the defined strain systems for many years. Using mixed strain starters rotations are essential and the total throughput of strains, many of which may be lysogens and a source of phages, is large in each factory. The complexity of these starters is such that meaningful phage relationships are difficult to determine and the operation of culture rotations relies heavily on experience. With defined strain systems, phage relationships may be determined with some accuracy providing a firm basis for culture rotations.

In New Zealand, for instance, pairs of strains were used on a rotational basis for many years, though a single culture of six carefully selected strains has performed satisfactorily for some time in several cheese factories (Limsowtin *et al.*, 1977; Lawrence *et al.*, 1978).

The strains used in the New Zealand starter system are selected after exposure to factory wheys, i.e. possess resistance to phages in those factory environments. The concept of using multiple strain starters composed of phage insensitive strains has now been extended by Daniell & Sandine (1981) and Thunell *et al.* (1981). Six-strain starters resistant to a bank of phages are used without rotation but wheys are monitored regularly and if phage appears against any of the strains, that strain is withdrawn and an alternative phage resistant mutant developed with which to replace it. In the interim, the starter is capable of continuing production without loss of activity. Thunell *et al.* (1981) showed that factories in the USA which changed to this system experienced a pronounced reduction in starter failure rate and produced a higher percentage of top graded cheese. More recently, a similar system has been adopted by some factories in Ireland (Daly, 1983a, 1983b) and they have reported a similar experience. Richardson *et al.* (1980) claim success using paired phage-insensitive mutants coupled with phage monitoring. In this instance, however, a refrigerated reserve culture is held for immediate replacement of strains against which phage might develop.

A novel approach suggested recently by Gamay *et al.* (1982) is the use of proteinase negative variants for cheesemaking. These are used at high inoculum levels to overcome their poor multiplication rate in milk. Their slow rate of multiplication does, however, prevent phages developing during incubation.

There seems little doubt that the move towards defined strain systems will continue; while they do not totally prevent phage development, they offer an appreciably greater measure of control. Furthermore, they permit the introduction of new or improved strains on a rational basis and if the potential of modern genetics techniques is to be realized, are an essential development.

6. CONCLUSIONS

Although phages of lactic-acid bacteria have been studied for over fifty years, they continue to cause problems in dairy fermentations. Much of the early research addressed these problems directly and achieved some measure of containment through the development of physically protected systems, improved disinfection and often empirically based starter rotations.

In recent years, however, there has been a greater emphasis on fundamental studies of the phages themselves and in the long term it is from this approach that a satisfactory solution to the phage problem is most likely to emerge.

Details of ultrastructure are becoming well documented though electron-micrographs of consistently high quality are still elusive and techniques for specimen preparation need close attention. Serological identification of phages has not been pursued extensively but further investigations of the sensitive ELISA systems (e.g. Lembke & Teuber, 1979) may be worthwhile. There is little information concerning receptor sites and the mechanisms of adsorption, penetration etc. for these phages. With few exceptions, e.g. studies of receptors in cell walls and membranes of lactic streptococci (Oram & Reiter, 1968) and of *Lb. casei* (Yokokura, 1977; Watanabe *et al.*, 1980), the area has been badly neglected and there is a clear need for work to be done. This is particularly so in view of the recently described resistance mechanism in lactic streptococci whereby plasmid DNA has been shown to encode an adsorption blocking mechanism (Sanders & Klaenhammer, 1983, 1984; de Vos *et al.*, 1984).

It is probably in the area of molecular biology that the most immediate advances will be made. Analysis of DNA restriction digests has already been shown to have taxonomic value (Davies *et al.*, 1981; Shimizu-Kadota *et al.*, 1983) and the prophage origin of a virulent phage of *Lb. casei* was established through a study of their DNA sequences (Shimizu-Kadota *et al.*, 1983). These approaches will undoubtedly be used to probe other virulent–temperate phage relationships in the future. Restriction–modification systems and the plasmid DNA encoding them will also be the subjects of analysis at the molecular level. Transfer of R/M systems (e.g. Gasson, 1984) and cloning of restriction enzymes in starter strains as suggested by Daly & Fitzgerald (1982) are probable developments. The potentially broad spectrum resistance conferred by plasmid DNA in some strains showing reduced phage/adsorption (Sanders & Klaenhammer, 1983; de Vos *et al.*, 1984) has also been shown to be transferable (de Vos, Davies & Underwood, unpublished data) and is likely to be exploited in strain development.

Thus investigations at the molecular level will shed light on the relationships and origins of phages and may eventually lead to the construction of starter strains with stable and broad spectrum phage resistance. In the meantime, however, measures for phage containment must be maintained and defined strain systems using phage insensitive mutants (e.g. Thunell *et al.*, 1981) probably offer the best approach to change phage control in practice.

REFERENCES

Accolas, J-P. & Spillmann, H. (1979a) *J. Appl. Bacteriol.* **47**, 135.
Accolas, J-P. & Spillmann, H. (1979b) *J. Appl. Bacteriol.* **47**, 309.
Bauer, H., Dentan, E. & Sozzi, T. (1970) *J. Microscopie* **9**, 891.

BIRLISON, F. G. T. & STANLEY, G. (1982) *Dairy Industries Int.* **47**, 27.
BOISSONNET, B., BOZIO, J. Y., BRIERE, C., LECLERC, M-C. & LARPENT, J. P. (1981) *Le Lait* **61**, 134
BOUSSEMAER, J. P., SCHRAUWEN, P. P., SOURROUILLE, J. P. & GUY, P. (1980) *J. Dairy Res.* **47**, 401.
BRADLEY, D. E. (1967) *Bacteriol. Rev.* **31**, 230.
CHAPMAN, H. R. (1978) *J. Soc. Dairy Technol.* **31**, 99.
CHOPIN, M-C. (1980) *J. Dairy Res.* **47**, 131.
CHOPIN, M-C. & ROUSSEAU, M. (1983) *Appl. Environ. Microbiol.* **45**, 294.
CHOPIN, M-C., CHOPIN, A. & ROUX, C. (1976) *Appl. Environ. Microbiol.* **32**, 741.
CHOPIN, M-C., ROUAULT, A. & ROUSSEAU, M. (1983) *Le Lait* **63**, 102.
COETZEE, J. N. & DEKLERK, H. C. (1962) *Nature* (London) **194**, 505.
COLLINS, E. B. (1956) *Virology* **2**, 261.
COX, W. A. & LEWIS, J. E. (1972) In *Safety in Microbiology*, Technical Series No. 6, Society for Applied Bacteriology. Academic Press, London, p. 133.
CZULAK, J., BANT, D. J., BLYTH, S. C. & GRACE, J. B. (1979) *Dairy Industries Int.* **44**, 17.
DALY, C. (1983a) *Irish J. Food Sci. Technol.* **7**, 39.
DALY, C. (1983b) *Antonie van Leeuwenhoek* **49**, 297.
DALY, C. & FITZGERALD, G. F. (1982) In *Microbiology 1982*. D. Schlessinger, Ed. American Society for Microbiology, Washington, DC.
DANIELL, S. D. & SANDINE, W. E. (1981) *J. Dairy Sci.* **64**, 407.
DAVIES, F. L. & GASSON, M. J. (1981) *J. Dairy Res.* **48**, 363.
DAVIES, F. L., UNDERWOOD, H. M. & GASSON, M. J. (1981) *J. Appl. Bacteriol.* **51**, 325.
DEANE, E., NELSON, F. E., RYSER, F. C. & CARR, P. H. (1953) *J. Dairy Sci.* **36**, 185.
DEKLERK, H. C. & HUGO, N. (1970) *J. General Virology* **8**(3), 231.
DELBRUCK, M. (1940) *J. General Physiol.* **23**, 643.
DENTAN, T., SOZZI, T. & BAUER, H. (1970) *J. Microscopie* **9**, 567.
DE VOS, W. M., UNDERWOOD, H. M. & DAVIES, F. L. (1984) *FEMS Microbiol. Lett.* **23**(2).
DHILLON, T. S., DHILLON, E. K. S. & LINN, S. (1982) *Canadian J. Microbiol.* **28**, 141.
DUCKWORTH, D. H., GLENN, J. & McCORQUODALE, D. J. (1981) *Microbiol. Rev.* **45**, 52.
ERICKSON, R. J. (1980) *Dairy Industries Int.* **45**, 37.
FITZGERALD, G. F., DALY, C., BROWN, L. R. & GINERAS, T. R. (1982) *Nucleic Acids Res.* **10**, 8171.
GAMAY, A. Y., HAFEZ, R. S., BROWN, R. J., ERNSTROM, C. A. & RICHARDSON, G. H. (1982) *J. Dairy Sci.* **65**, Suppl. 1, 51.
GASSON, M. J. (1984) *FEMS Microbiol. Lett.* **21**, 7.
GASSON, M. J. & DAVIES, F. L. (1980) *Appl. Environ. Microbiol.* **40**, 964.
GELIN, M., WURCH, T. & LINDER, R. (1970) *Comptes-rendus Hebdomaires des Seances de l'Academie des Sciences de Paris,* Series D, **270**, 425.
GEORGHIOU, D., PHUA, S. H. & TERZAGHI, E. (1981) *J. General Microbiol.* **122**, 295.

GUDKOV, A. V., OSTROUMOV, L. A. & ODEGOV, N. I. (1982) *Proc. XXI Int. Dairy Congress, Moscow,* Vol. 1, Book 1. Mir Publishers, Moscow, p. 311.

GULSTROM, T. J., PEARCE, W. E., SANDINE, W. E. & ELLIKER, P. R. (1979) *J. Dairy Sci.* **62,** 208.

HEAP, H. A. & JARVIS, A. W. (1980) *New Zealand J. Dairy Sci. Technol.* **15,** 75.

HEAP, H. A. & LAWRENCE, R. C. (1977) *New Zealand J. Dairy Sci. Technol.* **12,** 213.

HEAP, H. A., LIMSOWTIN, G. K. Y. & LAWRENCE, R. C. (1978) *New Zealand J. Dairy Sci. Technol.* **13,** 16.

HENNING, D. R., BLACK, C. H., SANDINE, W. E. & ELLIKER, P. R. (1968) *J. Dairy Sci.* **51,** 1.

HUGGINS, A. R. & SANDINE, W. E. (1977) *Appl. Environ. Microbiol.* **33,** 184.

HUGGINS, A. R. & SANDINE, W. E. (1979) *J. Dairy Sci.* **62,** 70.

HULL, R. R. (1977) *Australian J. Dairy Technol.* **32,** 65.

JARVIS, A. W. (1977) *New Zealand J. Dairy Sci. Technol.* **12,** 176.

JARVIS, A. W. (1978) *Appl. Environ. Microbiol.* **36,** 785.

JARVIS, A. W. (1981) *New Zealand J. Dairy Sci. Technol.* **16,** 25.

JARVIS, A. W. (1982) *Proc. XXI Int. Dairy Congress, Moscow,* Vol. 1, Book 2. Mir Publishers, Moscow, p. 314.

KLAENHAMMER, T. R. & MCKAY, L. L. (1976) *J. Dairy Sci.* **59,** 396.

KOZAK, W., RAJCHERT-TRZPIL, M., ZAJDEL, J. & DOBRZANSKI, W. T. (1973) *Appl. Microbiol.* **25,** 305.

KRINGELUM, B. W. (1982) *Proc. XXI Int. Dairy Congress, Moscow,* Vol. 1, Book 2. Mir Publishers, Moscow, p. 326.

KEOGH, B. P. & SHIMMIN, P. D. (1974) *Appl. Microbiol.* **27,** 411.

KING, W. R., COLLINS, E. B. & BARETT, E. L. (1983) *Appl. Environ. Microbiol.* **45,** 1481.

LAWRENCE, R. C. (1978) *New Zealand J. Dairy Sci. Technol.* **13,** 129.

LAWRENCE, R. C., THOMAS, T. D. & TERZAGHI, B. E. (1976) *J. Dairy Res.* **43,** 141.

LAWRENCE, R. C., HEAP, H. A., LIMSOWTIN, G. & JARVIS, A. W. (1978) *J. Dairy Sci.* **61,** 1181.

LEDFORD, R. A. & SPECK, M. L. (1979) *J. Dairy Sci.* **62,** 781.

LEMBKE, J. & TEUBI R, M. (1979) *Milchwissenschaft* **34,** 457.

LEMBKE, J. & TEUBER, M. (1981) *Milchwissenschaft* **36,** 10.

LEMBKE, J. & TEUBER, M. (1982) *XXI Int. Dairy Congress, Moscow* **1**(2), 334.

LEMBKE, J., KRUSCH, U., LOMPE, A. & TEUBER, M. (1980) *Zentralblatt fur Bakteriologie Mikrobiologie und Hygiene I. Abteilung Originale C1,* **79.**

LIMSOWTIN, G. K. Y. & TERZAGHI, B. E. (1976) *New Zealand J. Dairy Sci. Technol.* **11,** 251.

LIMSOWTIN, G. K. Y., HEAP, H. A. & LAWRENCE, R. C. (1977) *New Zealand J. Dairy Sci. Technol.* **12,** 101.

LIMSOWTIN, G. K. Y., HEAP, H. A. & LAWRENCE, R. C. (1978) *New Zealand J. Dairy Sci. Technol.* **13,** 1.

LOOF, M., LEMBKE, J. & TEUBER, M. (1983) *Systematic Appl. Microbiol.* **4,** 413.

LOWRIE, R. V. (1974) *Appl. Microbiol.* **27,** 210.
MARSHALL, R. J. & BERRIDGE, N. J. (1976) *J. Dairy Res.* **43,** 449.
MARTIN, M. (1983) *La Technique Laitiere* No. 976, 45.
MEISTER, K. A. & LEDFORD, R. A. (1979) *J. Food Protection* **42,** 396.
MULLAN, W. M. A. & CRAWFORD, R. J. M. (1981). *Irish J. Food Sci. Technol.* **5,** 24.
MULLAN, W. M. A. & CRAWFORD, R. J. M. (1982a) *Proc. XXI Int. Dairy Congress, Moscow,* Vol. 1, Book 2. Mir Publishers, Moscow, p. 350.
MULLAN, W. M. A. & CRAWFORD, R. J. M. (1982b) *Proc. XXI Int. Dairy Congress, Moscow,* Vol. 1, Book 2. Mir, Publishers, Moscow, p. 349.
NAYLOR, J. & CZULAK, J. (1956) *J. Dairy Res.* **26,** 126.
NYIENDO, J. A. (1974) PhD Thesis, Oregon State University.
OGATA, S. (1980) *Biotechnol. Bioeng.* **22,** Suppl. 1, 177.
ORAM, J. D. & REITER, B. (1965) *J. General Microbiol.* **40,** 57.
ORAM, J. D. & REITER, B. (1968) *J. General Virology* **3,** 103.
PEAKE, S. E. & STANLEY, G. (1978) *J. Appl. Bacteriol.* **44,** 321.
PEARCE, L. E. (1978) *New Zealand J. Dairy Sci. Technol.* **13,** 166.
PEARCE, L. E. & LOWRIE, R. J. (1974) *Proc. XIX Int. Dairy Congress, New Delhi,* **IE,** p. 410.
POP, M. & KURMANN, J. L. (1981) *Schweizerische Milchzeitung* **107,** 317.
PYATNITSYNA, I. N. & ZADOYANA, S. B. (1978) *Proc. XX Int. Congress, Paris,* **E,** p. 567.
REINBOLD, G. W., REDDY, M. S. & HAMMOND, E. G. (1982) *J. Food Protection* **45,** 119.
REITER, B. (1956) *Dairy Industries* **21,** 877.
REYROLLE, J., CHOPIN, M-C., LETELLIER, F. & NOVEL, G. (1982) *Appl. Environ. Microbiol.* **43,** 349.
RICHARDSON, G. H., CHENG, C. Y. & YOUNG, R. (1977) *J. Dairy Sci.* **60,** 378.
RICHARDSON, G. H., HONG, G. L. & ERNSTROM, C. A. (1980) *J. Dairy Sci.* **63,** 1981.
SAKURAI, T., TAKAHASHI, T. & ARAI, H. (1970) *Japanese J. Microbiol.* **14,** 333.
SANDERS, M. E. & KLAENHAMMER, T. R. (1980) *Appl. Environ. Microbiol.* **40,** 500.
SANDERS, M. E. & KLAENHAMMER, T. R. (1981) *Appl. Environ. Microbiol.* **42,** 944.
SANDERS, M. E. & KLAENHAMMER, T. R. (1983) *J. Dairy Sci.* **66,** Suppl. 1, 57.
SANDERS, M. E. & KLAENHAMMER, T. R. (1984) *Appl. Environ. Microbiol.* **46,** 1125.
SANDINE, W. E. & AYRES, J. W. (1981) US Patent 4,282,255.
SAXELIN, M-L., NURMIAHO, E-L. & SUNDMAN, V. (1979) *Canadian J. Microbiol.* **25,** 1182.
SHIMIZU-KADOTA, M. & SAKURAI, T. (1982) *Appl. Environ. Microbiol.* **43,** 1284.
SHIMIZU-KADOTA, M. & TSUCHIDA, N. (1984) *J. General Microbiol.* **130,** 423.

SHIMIZU-KADOTA, M., SAKURAI, T. & TSUCHIDA, N. (1983) *Appl. Environ. Microbiol.* **45**, 669.
SHIN, C. & SATO, Y. (1979a) *Japanese J. Zootech. Sci.* **50**, 419.
SHIN, C. & SATO, Y. (1979b) *Japanese J. Zootech. Sci.* **50**, 638.
SHIN, C. & SATO, Y. (1980) *Japanese J. Zootech. Sci.* **51**, 478.
SHIN, C. & SATO, Y. (1981) *Japanese J. Zootech. Sci.* **52**, 639.
SINHA, R. P. (1980) *Appl. Environ. Microbiol.* **40**, 326.
SOZZI, T. (1977) *Il Latte* **2**, 31.
SOZZI, T. & MARET, R. (1975) *Le Lait* **55**, 269.
SOZZI, T., POULIN, J. M., MARET, R. & POUSAZ, R. (1978) *J. Appl. Bacteriol.* **44**, 159.
SOZZI, T., BAUER, H., MARET, R. & DENTAN, E. (1980) *Milchwissenschaft* **35**, 17.
STETTER, K. O. (1977) *J. Virology* **24**, 685.
TAMINE, A. Y. (1981) In *Dairy Microbiology,* Vol. 2, R. K. Robinson, Ed. Applied Science Publishers Ltd, London.
TERZAGHI, B. E. (1976) *New Zealand J. Dairy Sci. Technol.* **11**, 155.
TERZAGHI, B. E. & SANDINE, W. (1981) *J. General Microbiol.* **122**, 305.
TEUBER, M. & LEMBKE, J. (1982) In *Microbiology 1982,* D. Schlessinger, Ed. American Society for Microbiology, Washington, DC.
TEUBER, M. & LEMBKE, J. (1983) *Antonie van Leeuwenhoek* **49**, 283.
THUNELL, R. K., SANDINE, W. E. & BODYFELT, F. W. (1981) *J. Dairy Sci.* **64**, 2270.
TSANEVA, K. P. (1976) *Appl. Environ. Microbiol.* **31**, 590.
VERHUE, W. M. (1978) *Appl. Environ. Microbiol.* **35**, 1145.
VON FROHLICH, G., DEGLE, I. & BUSSE, M. (1978) *Milchwissenschaft* **33**, 428.
VON KURMAN, J. L. (1979) *Schweizerische Milchwirtschaftliche Forschung* **8**, 71.
WALKER, A. L., MULLAN, W. M. A. & MUIR, M. E. (1981) *J. Soc. Dairy Technol.* **34**, 78.
WATANABE, K., TAKESUE, S., ISHIBASHI, K. & NAKAHARA, S. (1980) *Agric. Biol. Chem.* **44**, 869.
WIGLEY, R. C. (1980) *J. Soc. Dairy Technol.* **33**, 24.
WILLRETT, D. L., SANDINE, W. E. & AYRES, J. W. (1982) *Cultured Dairy Products J.* **17**, 5.
YOKOKURA, T. (1977) *J. General Microbiol.* **100**, 139.
YOKOKURA, T., KODAIRA, S., ISHWA, H. & SAKURAI, T. (1974) *J. General Microbiol.* **84**, 277.

Chapter 6

Flavour Development in Fermented Milks

VALERIE M. E. MARSHALL

National Institute for Research in Dairying, Shinfield, Reading, UK

1. INTRODUCTION

1.1. The Diversity of Fermented Milks

With the exception of cheese, yoghurt is the only fermented milk product consumed in any quantity in the UK. This is probably because the tradition for fermenting milk declined as the quality and quantity of fresh milk improved. In many countries, however, a variety of fermented milks are consumed, often in preference to fresh milk (Table 1). The fermented products may be more attractive to the consumer because of their characteristic properties—flavour, aroma, appearance and texture. Furthermore, they are safer than fluid milk in countries where transport, pasteurization and refrigeration facilities are inadequate. Fermentation is the oldest method for preserving milk and has been developed independently all over the world. As a result the organisms responsible for souring differ in the various countries. These differences are reflected in the types of cultured milk they produce. They can generally be divided into four broad categories based on the microbial species which dominate(s) the flora. Table 2 shows a relationship between the type of product, the microflora and the geographical distribution.

As early as 1900 Conn attached importance to bacterial growth in the development of aroma in cultured products. He concluded that cream ripening involved more than souring when he found that addition of acid did not accomplish the same results as bacterial growth. He suggested that acid and flavour were the results of different fermentations and that aroma was separate from flavour. The truth of these

153

TABLE 1
Consumption of Fermented Milks (kg/person/year) (figures from International
Dairy Federation, 1983)

Country	Buttermilk	Yoghurt	Acidophilus milk	Kefir and koumiss
Canada	0·4	1·8	—	—
Denmark	17·4	9·2	0·4	—
Finland	28·5	8·2	—	2·0
France	0·9	8·2	—	—
Israel	13·6	3·9	—	—
Japan	—	1·5	—	—
Sweden	17·2	3·6	0·8	2·8
W. Germany	4·9	5·9	0·5	—
U.K.	—	2·8	—	—
USSR	—	7·2	—	4·5
Bulgaria	—	31·5	—	—

TABLE 2
Relationship Between Type of Cultured Milk and Geographical Area

Type	Culture(s) used	Region
I	Streptococcal and leuconostoc species	Norway, Sweden, Finland, Iceland
II	Lactobacillus species	Bulgaria, Japan
III	Streptococcal and lactobacillus species	Egypt, Iraq, Lebanon, Syria, Turkey, India
IV	Streptococcal and lactobacillus species and yeasts	USSR, Lebanon, Finland

observations is still evident today when we make so careful a choice in blending mixed starter strains to bring out the best flavour and aroma in our fermented dairy products. Fermentation of milk is carried out by micro-organisms (yeasts, bacteria, moulds) which are capable of fermenting lactose predominantly to lactic acid. This is responsible for the sharp refreshing taste of all fermented milks and although non-volatile it serves as an excellent background for the more distinctive flavours and aromas characteristic of each fermented milk. Although the predominant metabolite is lactic acid it is clear that the minor metabolites are vital for product quality and identity. Also important is a cultured milk of good consistency, and organisms producing polymers of protein (ropy strains) may be added to improve texture, or those producing polymers of carbohydrate (glucose–galactose) may be necessary for starter integrity, e.g. kefiran and kefir grain formation. Consistency may be destroyed by gas production if fermentation is heterofermentative, but the presence of carbon dioxide is a requirement for the alcoholic milk beverages, kefir and koumiss, as it imparts a sparkling character which turns these products into refreshing summer drinks. All these organoleptic qualities are consequences of multiple fermentations and it is accepted that selection of the multi-strain or more often multi-species inoculum is necessary to obtain a good fermented product.

1.2. Starter Organisms and their Metabolites

A hundred years ago starter cultures were unknown, but throughout this century dairy microbiologists have unravelled the mixes of organisms responsible for much of the flavour and aroma. Table 3 lists the

TABLE 3
Starter Cultures for Milk Fermentation and their Principal Metabolic Products

Starter organisms	*Important metabolic products*
Mesophilic bacteria	
Str. cremoris	lactate
Str. lactis	lactate
Str. lactis subsp. *diacetylactis*	lactate, diacetyl
Leuc. cremoris[a]	lactate, diacetyl
Lb. acidophilus	lactate
Lb. brevis	CO_2, acetate, lactate
Lb. casei	lactate
Thermophilic bacteria	
Str. thermophilus	lactate, acetaldehyde
Lb. bulgaricus	lactate, acetaldehyde
Lb. lactis	lactate
Yeasts	
Saccharomyces cerevisiae	ethanol, CO_2
Candida (Torula) kefir	ethanol, CO_2
Kluveromyces fragilis	acetaldehyde, CO_2

[a] *Leuc. cremoris* = *Leuc. citrovorum* = *Leuc. mesenteroides* subsp. *cremoris*.

many organisms which have been isolated from various milks together with their metabolic products. This gives a clue to their role in flavour production. Some of the starters are mesophilic, others are thermophilic. For further characterization and for more information about distinguishing features the reader is referred to Bergey (1974) and Lodder (1970).

Organisms are selected and combined depending on their use. For example fast acid-producing, thermophilic organisms are used for yoghurt manufacture; buttermilk requires a slower rate of acid production, and for kefir a gentle release of carbon dioxide at low temperatures is necessary. Starters of mixed species are now frequently available commercially. But analysis of growth, balance and metabolic activity is essential if the products are to be manufactured with consistent good quality. Standards for propagation by today's starter manufacturers need to be high. Some starters are more sensitive to poor subculturing than others. Petterssen (1975), for example, found that lactic acid production by mixed buttermilk cultures was poor if incubation was too long prior to harvesting and making the bulk starter. Lactic acid accumulated most rapidly during the exponential phase of growth and further incubation resulted in loss of the streptococcus relative to the leuconostoc. Sensitivity to lactic acid of some species of streptococcus has also been reported by Bergère & Hermier (1968) and if mixed starters are to contain these then attention must be given to the time of subculturing and harvesting when preparing starters to ensure that the correct balance is maintained. Yoghurt starters would appear to be more stable to changes of balance. Quite different ratios of rods : cocci can be tolerated, the unbalanced culture soon makes up the deficit (Accolas *et al.*, 1977), and Berridge (1966) notes a self-regulatory capacity of a mixed streptococcal starter when grown in continuous culture. Methods of propagation are therefore important.

To keep and maintain a good starter for a particular fermented product it is essential to know what is expected of it in terms of flavour and aroma. This knowledge is not always available and many products are poor because of this. Knowledge of the biochemical pathways leading to flavour production can help in making the right choice of starter. This is the area in which the dairy microbiologist can make a great contribution. This chapter attempts to cite some examples of what can be expected of a given organism in terms of its ability to produce flavour when grown in milk. An excellent review on the biosynthesis of flavour compounds by micro-organisms has been compiled by Collins (1972) which clarifies many of the pathways leading to important end metabolites.

2. METABOLIC ROUTES TO FLAVOUR PRODUCTION

2.1. Catabolism and Anabolism

The utilization of food materials by any living organism proceeds by specific sequences of enzymic reactions which collectively are described as metabolic pathways. These fulfil two main functions: they supply the energy necessary for (a) biosynthetic and (b) other energy-dependent processes. Pathways giving rise to precursors are catabolic, those which are synthetic are anabolic and there are central pathways which link the two. Catabolism is essentially oxidative and therefore produces reducing power. Much of the reducing power generated during catabolism is consumed during anabolism. The reducing power is provided by pyridine nucleotides (NAD or NADP) which are very limited and have to be recycled. The availability of this reducing power dictates the metabolic products. Fermentation products are reduced organic carbon compounds formed from re-oxidizing reduced pyridine nucleotides. As a general rule the further an organism can reduce a product the better, the limitation is the battery of enzymes with which it is endowed. The organism reduces organic carbon to the limit of its ability and then excretes the end metabolites to prevent any possibility of re-oxidation within the cell. These end metabolites are often volatile and will therefore contribute to the flavour and aroma of the fermented product. Those with which we are most concerned are acetate, acetaldehyde, diacetyl, ethanol and acetoin. It is only during the last fifteen years that studies have been conducted concerning the pathways leading to the excretion of these metabolites by the starter organisms concerned in the manufacture of fermented milks and not all starters have been under close investigation.

Figure 1 shows many of the relevant catabolic pathways concerned with sugar breakdown, but it is important to realize that not all of the pathways are common to all of the micro-organisms which ferment milk. There are intra-species as well as inter-species differences. *Streptococcus cremoris* and *Str. lactis* generally have the tagatose pathway in addition to the Embden–Meyerhof pathway but some variants of *Str. lactis* may also use the Leloir pathway if galactose is present in the

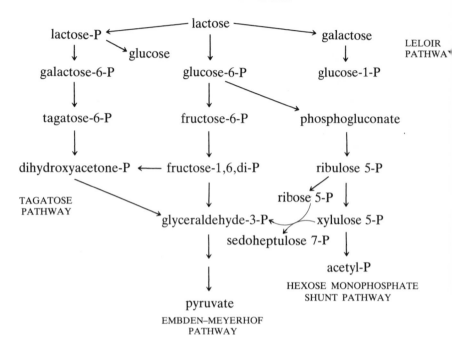

Fig. 1. Pathways of lactose catabolism.

growth medium (Thompson & Thomas, 1977; Thompson, 1978). *Str. thermophilus* appears to use only the Embden–Meyerhof pathway (O'Leary & Woychick, 1976; Tinson *et al.*, 1982). Premi *et al.* (1972) have shown the presence of β-D-phosphogalactosidase in a number of dairy lactobacilli. This enzyme produces intracellular galactose-6-phosphate which may then be metabolized by the tagatose pathway. Leuconostocs and lactobacilli are also capable of metabolizing glucose via the hexose monophosphate shunt, thus conserving energy because less ATP is consumed. Whatever the route of carbohydrate catabolism the important products are pyruvate, acetyl coenzyme A and acetyl phosphate (Collins & Bruhn, 1970). These are the key intermediates and it is their subsequent metabolic routes which are of interest. Each starter has its own way of metabolizing these and producing flavoursome metabolites. Where known, each will be described in subsequent sections. However, three examples of changes in end product metabolism will be discussed here to illustrate how differences in flavour can

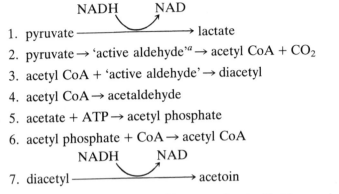

1. pyruvate $\xrightarrow{\text{NADH} \quad \text{NAD}}$ lactate
2. pyruvate → 'active aldehyde'^{a} → acetyl CoA + CO_2
3. acetyl CoA + 'active aldehyde' → diacetyl
4. acetyl CoA → acetaldehyde
5. acetate + ATP → acetyl phosphate
6. acetyl phosphate + CoA → acetyl CoA
7. diacetyl $\xrightarrow{\text{NADH} \quad \text{NAD}}$ acetoin

^{a} Thiamine pyrophosphate and lipoic acid are associated with this complex.

Fig. 2. Metabolism of pyruvate.

arise. The pathways describing the fates of pyruvate are shown in Fig. 2, reactions 1 to 5.

2.1.1. Changes in End Product Metabolism in Different Growth Phases
Catabolism of lactose to pyruvate by the Embden–Meyerhof pathway produces sufficient reduced nucleotides (NADH) to reduce pyruvate to lactate (reaction 1, Fig. 2). When NADH is limiting, perhaps at the end of the exponential growth phase or when growth is limited by other nutrients, such as amino acid availability, then pyruvate may not always be reduced to lactate because the organism cannot complete the pathway. Thus de Mann & Galesloot (1962) reported that actively growing leuconostoc produced little diacetyl (an alternative fate of pyruvate) but when NADH becomes unavailable as they enter the stationary phase, pyruvate is converted to diacetyl via acetyl CoA (reactions 2 and 3, Fig. 2) and excreted. These reactions do not involve a pyridine nucleotide.

2.1.2. Consequences of Increased Intracellular Intermediate Metabolites
Another way of stressing metabolism is to increase the intracellular pools of the important intermediates. Gilliland *et al.* (1970) showed that *Leuconostoc cremoris* (now renamed *Leuc. mesenteroides* subsp. *cremoris*, Garvie, 1983) produced diacetyl in broth only when sodium citrate was present. The pathway of citrate metabolism in this bacter-

ium is as follows:

$$8. \quad \text{citrate} \rightarrow \text{oxaloacetate} + \text{acetate}$$

$$9. \quad \text{oxaloacetate} \rightarrow \text{pyruvate} + CO_2$$

The enzyme responsible for reaction 8 is citrate lyase which is present in both the leuconostoc and *Str. lactis* subsp. *diacetylactis* (Harvey & Collins, 1963). Unlike glucose metabolism by the Embden–Meyerhof pathway there is no production of NADH so that while the pyruvate pool is elevated the reducing power is not. Conversion of pyruvate to diacetyl must then follow (reactions 2 and 3, Fig. 2) as NADH is not required for these pathways (Harvey & Collins, 1963; Keenan & Bills, 1968).

Acetaldehyde has been considered to have a central role in metabolism (Lees & Jago, 1976). Some starters produce this compound in small quantity, e.g. *Str. cremoris* and *Str. lactis* (Lindsay *et al.*, 1965), others in large quantities, e.g. *Str. thermophilus* (Pette & Lolkema, 1950b), and yet others, e.g. *Leuc. cremoris*, utilize it and are even stimulated by it (Keenan & Lindsay, 1966). Acetaldehyde is vital for yoghurt flavour (Pette & Lolkema, 1950c; Keenan & Bills, 1968; Bottazzi & Vescovo, 1969), but many starter organisms metabolize acetaldehyde to ethanol. The lack of alcohol dehydrogenase, the enzyme catalysing this reaction, in the two yoghurt starters (Lees & Jago, 1976) means that acetaldehyde is excreted as the end metabolite. To make a product with yoghurt flavour therefore requires starters with low, or absence of, alcohol dehydrogenase. *Lb. acidophilus* is not such an organism, but its metabolism can be stressed in much the same way as *Leuc. cremoris* is in the presence of citrate. Addition of threonine results in its being converted to glycine within the cell as follows:

$$10. \quad \text{threonine} \rightarrow \text{glycine} + \text{acetaldehyde}$$

$$11. \quad \text{acetaldehyde} \longrightarrow \text{ethanol}$$
$$\qquad\qquad\quad \text{NADH} \qquad \text{NAD}$$

The acetaldehyde pool is increased as a consequence of threonine metabolism but NADH is insufficient to reduce the acetaldehyde. The *Lb. acidophilus* therefore excretes the excess acetaldehyde (Marshall & Cole, 1983).

2.1.3. *Consequences of Genetic Change*

Mutations may also force metabolic pathways in other directions. Kuila & Ranganathan (1978) studied the acid and diacetyl production in a number of mutants of *Str. lactis* subsp. *diacetylactis*. Two types of mutant were identified: one which produced a titratable acidity of 0·39% and 138 ppm diacetyl, and the second mutant produced a titratable acidity of 0·93% and only 42·5 ppm diacetyl. Both mutants arose from a parent strain having intermediate values for titratable acidity and diacetyl. The restricted acid production (i.e. lactic acid) in the first type of mutant pushed the metabolism to diacetyl production in order to prevent pyruvate accumulation. McKay & Baldwin (1974) have also observed a large increase in flavour compounds by a spontaneous mutant of *Str. lactis* which was deficient in lactate dehydrogenase, the enzyme catalysing conversion of pyruvate to lactate (reaction 1, Fig. 2), and pyruvate had to be metabolized into other products. These examples demonstrate that flavour compounds are the products of necessity. The recurrent theme is the balance of oxidized and reduced pyridine nucleotides.

2.2. Inter-relationships Between the Starter Organisms

The symbiotic relationship between the yoghurt bacteria was reported by Orla-Jensen in 1931, by Pette & Lolkema in 1950 (a, b) and by Galesloot *et al.* (1968). In summary: *Str. thermophilus* is stimulated by free amino acids and peptides liberated from the milk proteins by *Lb. bulgaricus* (Miller & Kandler, 1964; Shankar & Davies, 1977). The acid production by *Lb. bulgaricus* is stimulated by a compound produced by *Str. thermophilus* in the absence of oxygen or at low oxygen concentration. This stimulus is formic acid (Veringa *et al.*, 1968). This may not be the only stimulant; production of CO_2 by the *Str. thermophilus* has also been implicated (Driessen *et al.*, 1982). The relationship between the two starter organisms may not be truly symbiotic. Symbiosis results in greater activity by mixed culture than the sum of the individual cultures. Accolas *et al.* (1977) studied a number of yoghurt cultures and demonstrated symbiosis in only one group, and this increase was not large (about 10%). If acid production was good in both *Lb. bulgaricus* and *Str. thermophilus*, then mixing the two offered no advantage. However, other work has shown substantial increases in acetaldehyde: *Str. thermophilus* produced a maximum of 4 ppm in 8 h and the *Lb. bulgaricus* produced 10 ppm in 8 h. Together they produced 23 ppm (Hamdan *et al.*, 1971). Driessen (1981) has termed the relationship proto-cooperation because the interaction is not obligatory (Odum, 1971). Thus each population produces a substance not initially present which influences the growth of the other.

Inter-relationships between multi-species are not easily studied as Driessen (1981) has pointed out. The components of the microflora often have similar metabolism, a minor part of which is complementary. The influence of the stimulatory compounds is difficult to describe theoretically, and practically is subject to change. For instance, taking the specific case of the two yoghurt starters, if for any reason there is a decrease in nitrogen availability for the *Str. thermophilus* a smaller growth rate of the organism follows and consequently a decrease in production of formic acid and CO_2 with less stimulation of the *Lb. bulgaricus*. This will break down less protein and still less nitrogen will be available and so on. Accolas *et al.* (1977), however, have observed that yoghurt starter is relatively stable to 10-fold changes in balance and a model for an inter-relationship which includes inhibition of each population has been proposed (Ubbels, 1981) and can be supported by evidence from Moon & Reinbold (1976) who demonstrated that *Str. thermophilus* can inhibit *Lb. bulgaricus* as the mixed culture enters the stationary phase. The models are only applicable to studies on continuous culture, but a shift of metabolism as a culture enters different stages of growth is not only important to flavour production by one population, but it also influences flavour production by its associative population. In buttermilk cultures, for instance, production of diacetyl by the leuconostocs is favoured by low pH (Gilliland *et al.*, 1970; Petterssen, 1975) when growth of the organism is slowed, pH is lowered as a consequence of lactate production during the exponential growth of the starter and its accumulation results in a change of end metabolites.

The behaviour of mixed starters and their influence on one another is clearly complex. Yoghurt, the best studied, contains only two populations; kefir offers a greater challenge, containing up to five different types of organism with the added intrigue of a mix of eukaryotic (yeast) and prokaryotic (bacterial) cells.

3. FERMENTED MILKS OF TYPE I

In this type of milk streptococcal species predominate, but the contribution of the leuconostoc is also important. The starters are available commercially and contain *Str. lactis* subsp. *diacetylactis*. The metabolic pathways leading to flavour production by these organisms are described in Fig. 2.

3.1. Cultured Buttermilk

This is consumed in quantity in the USA and the Irish Republic. It is a pourable, drinkable product which is taken as a thirst-quencher in summer. Traditionally buttermilk was the by-product of farmhouse buttermaking after churning cream ripened with *Str. cremoris* or *lactis* plus leuconostoc species and there is still a demand for a product which contains butterflakes (Kosikowski, 1977). Variations in the churning process and differences in the quality of cream during the year meant that a high quality product was not always available and to supply the demand for this beverage cultured buttermilk was made using skimmilk.

Today's method of making cultured buttermilk involves pasteurization, inoculation with a commercial starter and incubation at 22°C to a titratable acidity of 0·7–0·9% (pH 4·6–4·5), which takes about 14–16 h. The flavour is sharp and buttery as a consequence of lactic acid and diacetyl production. Salt (0·1%) may be added.

3.1.1. Starter Composition and Diacetyl Production

Citrate can be utilized by *Str. lactis* subsp. *diacetylactis* and by *Leuc. cremoris* (reactions 8 and 9, p. 160) and the pyruvate metabolized according to Fig. 2 to produce more diacetyl. Fresh milk from cows on pasture contains 0·17% citrate but this declines during winter feeding (Marier & Boulet, 1958; Hegazi & Abo-Elnaga, 1980). Thus fortification of milk with citric acid or sodium citrate will improve flavour. Buttermilk may have a flat or insipid flavour if diacetyl is not produced.

In addition to *Str. lactis* and/or *cremoris*, modern commercial starters may include *Str. lactis* subsp. *diacetylactis* which, under aerobic conditions, will produce up to 47 μmoles acetoin + diacetyl per mg dry weight of cells (Collins & Bruhn, 1970). Unlike the leuconostocs however, this organism, in common with the other starter streptococci, also produces acetaldehyde (reaction 4, Fig. 2) which in excess produces an undesirable flavour in buttermilk described as 'green' (Keenan *et al.*, 1966). The inclusion of *Leuc. cremoris* in the starter is advantageous because not only does it not produce acetaldehyde, it can scavenge it (Lindsay *et al.*, 1966).

Acetaldehyde can undergo reduction or oxidation to produce two other end metabolites:

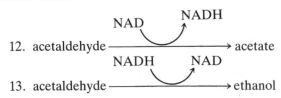

12. acetaldehyde ⟶ acetate

13. acetaldehyde ⟶ ethanol

These reactions occur in resting cells of *Str. lactis* subsp. *diacetylactis* and in *Leuc. cremoris* (Lees & Jago, 1978b), but in growing cells there is a need to regenerate NAD for glucose degradation and reaction 13 predominates (Speckman & Collins, 1968) but when the NADH is not required (resting cells) both reactions can occur.

3.1.2. Loss of Flavour in Buttermilk

Lack of the diacetyl component may occur in spite of the correct starter formulation. Pyruvate may be metabolized via 'active aldehyde' to acetoin (reactions 14 and 15 below). Acetaldehyde when complexed with thiamine pyrophosphate to form 'active aldehyde' is not a flavour volatile, nor is acetoin (Speckman & Collins, 1968).

14. pyruvate + 'active aldehyde' → acetolactate

15. acetolactate → acetoin + CO_2

16. acetoin ⟶ 2,3,butylene glycol

If NADH is available then reaction 16 and reaction 7 (Fig. 2) can also occur.

Conversion of pyruvate to acetyl CoA (reaction 2, Fig. 2) is an important step in flavour production and this reaction can be inhibited in *Str. lactis* and *Str. cremoris* by unsaturated fatty acids (Anders & Jago, 1970). Thus diacetyl formation may be hindered if full-fat milk is used for buttermilk manufacture. Leuconostocs, however, can make acetyl CoA by a route which does not involve pyruvate. They can ferment glucose via the hexose monophosphate shunt and cleave xylulose phosphate to acetyl phosphate (Fig. 1) which is converted to acetyl CoA. Thus acetyl CoA formation does not in this case involve pyruvate and the pathway may provide an added route to flavour formation.

3.1.3. Texture of Buttermilk

Viscosity contributes to organoleptic quality. Fluidity is desirable in buttermilk; it is a fermented milk which is taken as a thirst-quencher, but too little viscosity is almost as objectionable as too much. It must linger in the mouth long enough for the flavour volatiles to be recognized and appreciated. A buttermilk which is too thin is often associated with too low an acidity (high pH) when the curd is broken, the critical pH at which the curd should not be disturbed is 4·7–5·0. Too much agitation during incubation can lead to loss of texture and gel disruption can occur from excess CO_2 production giving rise to the phenomenon of 'curd floating'. It is clear that the metabolism of pyruvate to compounds other than lactate gives rise to CO_2 (reaction 2, Fig. 2). It is therefore important to keep this to a minimum. The correct balance of cultures is therefore required. This balance has been investigated by Petterssen (1975) for the Scandinavian buttermilks and a 20% proportion of aroma bacteria has been suggested.

3.2. Scandinavian Buttermilks

These milks are popular in Norway and Sweden; filmjolk, lattfil and langfil (meaning long milk) being the most common. The Swedes consume 17·2 kg buttermilk/person/year (Table 1). The largest sales are of filmjolk taking nearly 68% of the market in 1975 (Bertelsol & Jonsson, 1976). Most of this is consumed in place of fresh milk—over cereals or on its own as a beverage. Filmjolk contains 3% fat and lattfil only 0·5%. Both are fermented with starters containing *Str. cremoris* and/or *Str. lactis, Str. lactis* subsp. *diacetylactis* and *Leuc. cremoris* in much the same way as the buttermilk described above. Filmjolk with its 5% fat has a sharp, creamy, full flavour, good consistency and an almost glossy look. Acetaldehyde and diacetyl both contribute to the aroma and the lactic acid takes the pH to 4·4–4·6. Lattfil is very similar but lacks the creaminess of its full fat cousin. Langfil is made with a multistrain starter which includes a strain of streptococcus which produces 'ropiness'. Before becoming commercially available langfil was made locally on farms and tradition has it that the inoculum was enriched by rubbing the interior of milk pails with leaves of the butterwort (*Pinguicla vulgaris*). This, it was thought, introduced to the milk a slime-producing bacterium whose general habitat was the leaf of this bog weed. A slime-producing bacterium (*Alcaligenes viscosum*) is associated with this plant but it is not used today to produce langfil (Nilsson & Nilsson, 1958). Commercially it is made using *Str. cremoris,* a ropy variant of *Str. lactis* and *Str. lactis* subsp. *diacetylactis*. The milk is incubated at lower temperatures (18–20°C) than for other buttermilks to ensure production of the slime. Analysis of the slime showed that it was not a polysaccharide but a protein (Sundman, 1953). The final pH is 4·2–4·4. The predominant organoleptic character is the ropiness: a good langfil is not easy to spoon. It is because of this ropiness that the product is fermented in the carton in which it is sold; it cannot be piped.

3.3. Ymer

Ymer, the Danish equivalent of filmjolk, contains 3% fat, 11% solids-not-fat, has a pH 4·4–4·6 and is pourable. It is made using a mixed starter containing *Str. cremoris* and *Str. lactis* subsp. *diacetylactis* (Delaney, 1977). The flavour volatiles are therefore diacetyl and acetaldehyde. Ymer may also be thickened by allowing the whey to drain off and then dressing the 'cheese' with 5% cream (Kosikowski, 1978). A similar product to this is made in Iceland where ewes' milk is used. It is very old, dating from before the tenth century and is called *Skyr*.

3.4. Villi

This is Finland's buttermilk. It is made in a similar way to the others but is considered superior because it has an added attraction, the surface growth of a mould (*Geotrichum candidum*). The presence of the mould is essential, giving a slight, white, granular coat, scarcely amounting to a crust. The flavour is similar to the Scandinavian filmjolk with a slight 'musty' aroma, attributable to the mould. The mould is said to be of great importance in promoting flavour formation by preventing oxidation (Lang, 1980), but as diacetyl production is favoured by aeration (Collins, 1972) the importance of the mould to this product has still to be determined.

4. FERMENTED MILKS OF TYPE II

4.1. Bulgarian Buttermilk

Very little information is available about this product. *Lb. bulgaricus* is used to ferment pasteurized whole milk. In view of the lack of formate in milk which has not been given a heat treatment in excess of 120°C (Miller *et al.*, 1964; Galesloot *et al.*, 1968; Shankar, 1977) it can be expected that fermentation is longer than for yoghurt fermentation when this organism is combined with *Str. thermophilus.*

An overnight incubation at 40–42°C would produce a highly acid product with a good 'clean' flavour and an aroma similar to that of yoghurt. This aroma is due to acetaldehyde which the *Lb. bulgaricus* can produce in large quantity in single culture (Hamdan *et al.*, 1971; Bottazzi *et al.*, 1973). Although acetaldehyde is produced from pyruvate it is often contained in a complex with thiamine pyrophosphate as active aldehyde which is further metabolized. Acetaldehyde can arise from other routes (Lees & Jago, 1978a,b) one of which is important in *Lb. bulgaricus*: the amino acid, threonine, can be metabolized to acetaldehyde as follows:

17. threonine → glycine + acetaldehyde

and this reaction may be more important for flavour production in this buttermilk than metabolism of pyruvate.

4.2. Yakult

This is a milk fermented in Japan by the Yakult Central Institute for Microbiological Research, Tokyo using a special variant of *Lb. casei*; it is delivered door-to-door by some 50 000 enthusiastic saleswomen on bicycles (Lang, 1980). The Japanese market for this product has been built partly on account of its health-giving benefits (Mada, 1981), and the culture may be combined with bifidobacteria (Mada, 1982). Skim milk plus sugar, or starch syrup, is fermented to the desired acidity, then flavoured with fruit and essences. The culturing produces minor constituents of citric, succinic, malic and acetic acids and the flavour volatiles acetaldehyde, diacetyl and acetone (Manus, 1979). Cardenas *et al.* (1980) have studied the metabolism of *Lb. casei* in more detail and have examined the production of flavour compounds from pyruvate. The presence of pyruvate in the growth medium promoted a large excretion of acetoin and diacetyl (Fig. 2). This finding of increased flavour in the presence of added pyruvate may point to ways of improving flavour in fermented milks which use *Lb. casei*.

4.3. Acidophilus Milk

Fermentation with *Lb. acidophilus* is commonly advocated because of its alleged health benefits (Speck, 1976; Gilliland *et al.*, 1978; Ayebo *et al.*, 1980; Alm, 1981). However, *Lb. acidophilus* (particularly strains isolated from the intestinal tract) grows slowly in milk, multiplying only five fold in 18–24 h (Fowler, 1969). The flavour and consistency of milk fermented with this organism is often poor. For this reason it has been incorporated into mixed starters used for yoghurt manufacture which are used to produce the flavour and aroma compounds (Rasic & Kurmann, 1978). There is, however, some doubt about the number of viable cells of *Lb. acidophilus* present in these products when they reach the consumer, as Gilliland & Speck (1977) have shown that the organism quickly dies out in the presence of *Str. thermophilus* and *Lb. bulgaricus.*

A non-fermented product known as 'sweet acidophilus milk' (Speck, 1975) contains large numbers of viable *Lb. acidophilus* added as frozen concentrate to pasteurized milk. This is a cold-shelf product and has the flavour and aroma of pasteurized milk. There are, however, ways of making a well-flavoured fermented milk using *Lb. acidophilus* only. A large Swedish dairy includes yeast extract (0·1%) in the milk to be fermented and achieves good acidity with little impairment of quality due to the presence of the yeast extract, but aroma is poor. Marshall *et al.* (1982) fortified milk with whey protein and obtained a product of pH 4·6 and acetaldehyde levels of 15 ppm. Further improvements to flavour were made by including threonine in the milk and stressing the metabolism by increasing the intracellular pool of acetaldehyde (reaction 17). *Lb. acidophilus* normally metabolizes acetaldehyde to alcohol and regenerates more NAD (reaction 13, p. 164) because it possesses an alcohol dehydrogenase (Marshall *et al.,* 1982). The fermented milk made using *Lb. bulgaricus* has high levels of acetaldehyde (up to 35 ppm) because the organism lacks this enzyme. By providing excess threonine for the metabolism of *Lb. acidophilus*, there is an increase in intracellular acetaldehyde but no additional increase in NADH. The alcohol dehydrogenase is limited by low concentration of NADH and cannot therefore reduce the acetaldehyde to ethanol. The organism is thus obliged to excrete the excess acetaldehyde and the product made from milk containing threonine and whey protein has good flavour and aroma similar to that of *Lb. bulgaricus* fermented milk.

Lb. acidophilus will also utilize pyruvate in the presence of glucose and produce diacetyl (Cardenas *et al.,* 1980). If pyruvate were to be included in the formulation, a buttery flavour note may be expected.

Other benign, natural inhabitants of the gastrointestinal tract, however, may also be capable of fermenting milk. Bifidobacteria are frequently isolated from the gastrointestinal tract and their importance stressed in establishing a stable gut microflora for the new-born infant (Gyllenberg & Roine, 1957; Haenel, 1970). Thus in addition to *Lb. acidophilus* fermented milk, Marshall *et al.* (1982) have also fermented milk with *Bifidobacterium bifidum*, *B. adolescentis*, *B. infantis* and *B. longum* by supplementing milk with whey proteins and adding threonine to stimulate acetaldehyde production.

5. FERMENTED MILKS OF TYPE III

In this type of fermentation both lactobacillus and streptococcal species are used. The organisms are thermophilic and fermentation is rapid.

5.1. Yoghurt

The most widely distributed and most studied of the fermented milks, yoghurt is today almost always made commercially with equal ratio of *Str. thermophilus* and *Lb. bulgaricus*. Many reviews and a book have been devoted to this one fermented milk (Humphries & Plunkett, 1969; Rasic & Kurmann, 1978; Tamime & Deeth, 1980; Deeth & Tamime, 1981) and these provide excellent sources of information. In keeping with the title of this chapter, however, only the biochemistry of flavour production by the starter will be explored. Yoghurt has a sharp, refreshing acid taste, the pH is usually 4·0–4·4 (Kosikowski, 1977). The relationship between the two starter organisms makes the flavour quite unlike that encountered in any other dairy food. Acetaldehyde, diacetyl, acetoin, acetone and butan-2-one are all present in yoghurt, but acetaldehyde is recognized as the chief flavour component (Pette & Lolkema, 1950c; Schultz & Hingst, 1954; Gorner *et al.*, 1968). It is unlikely that acetone is derived from the activity of starter bacteria (Keenan *et al.*, 1966), both this compound and butan-2-one are milk components (Rasic & Kurmann, 1978). Typical yoghurt flavour has been described as being similar to walnuts (Kosikowski, 1977). Optimum flavour and aroma is obtained between 23 and 41 ppm acetaldehyde (Gorner *et al.*, 1968). Diacetyl and acetoin result from metabolic activity of *Str. thermophilus* and are very low, rarely reaching 0·5 ppm. The presence of diacetyl contributes to the delicate, full flavour and aroma of yoghurt and is especially important if acetaldehyde is low because it can enhance yoghurt flavour (Groux, 1973). The absence of alcohol dehydrogenase activity in both *Str. thermophilus* and *Lb. bulgaricus* makes these starters incapable of metabolizing acetaldehyde to ethanol (Lees & Jago, 1977). The streptococcus may form acetaldehyde from glucose via pyruvate (Fig. 2) but according to Lees & Jago (1978a) only trace amounts are formed via this route by *Lb. bulgaricus*. Lees & Jago (1976) and Shankar (1977) suggest that threonine metabolism is responsible for the acetaldehyde accumulation (reaction 17, p. 167) and Lees & Jago (1978b) further suggest that the glycine by-product would stimulate growth of *Str. thermophilus*. Reports that *Lb. bulgaricus* is the species largely responsible for the formation of acetaldehyde in yoghurt are numerous (Gorner *et al.*, 1968; Pette & Lolkema, 1950c; Schultz & Hingst, 1954) and although both *Lb. bulgaricus* and *Str. thermophilus* are able to convert threonine to glycine because they possess the relevant enzyme, threonine aldolase (Lees & Jago, 1976), and both have been shown to produce acetaldehyde in single culture (Shankar, 1977; Marshall *et al.*, 1982), the contribution of each population to flavour is not easily discerned.

Growth of the two organisms has been studied in the chemostat (Driessen, 1981). *Lb. bulgaricus* grown in continuous cultures in the presence of sodium formate was washed out of the chemostat and further experiments demonstrated a requirement for CO_2. CO_2 is produced by *Str. thermophilus* in making acetyl CoA from pyruvate (reaction 2, Fig. 2), but formic acid may be released instead of CO_2 if acetyl phosphate is involved (Mahler & Cordes, 1971). Both these reactions are important to mixed starter growth.

As yoghurt is made in many different countries so the taste will differ. For instance, acid production by yoghurt starters has been reported to be faster in buffalo milk than in cows' milk (Khana & Singh, 1979) and this may also be true for goats' milk (Duitschaever, 1978). Goats' milk yoghurt, however, has less aroma than cows' milk yoghurt, having only 5 ppm acetaldehyde present (Abrahamsen, 1978). Some studies have been conducted on the contribution of volatile fatty acids to flavour. Caproic, butyric and propionic acids are all present in goats', sheep's and cows' milks and in yoghurts made from them, but their proportions differ and change during fermentation (Rasic & Kurmann, 1978). Their contribution to flavour will be small, however, as they are much less volatile than acetaldehyde and diacetyl.

5.2. Dahi

This is the Indian equivalent of yoghurt. It is not made commercially on a large scale and therefore its microbiology is extremely variable. It follows that taste and appearance are also variable. Made in mud pots, usually from buffalo milk, sometimes cows' milk, the bacteria used are *Lb. bulgaricus* and/or *Lb. plantarum* and *Str. thermophilus* and/or *Str. lactis*. The inoculum is higher than is usual for fermented milks (approximately 20%) and incubation is for 1–3 h only. As a consequence the pH is quite high (Bhattacharya *et al.*, 1980). The product made from buffalo milk has a more granular structure and consistency (Rasic & Kurmann, 1978, p. 97). *Lb. plantarum* is a heterofermentative lactobacillus, producing acetate in addition to lactate.

6. FERMENTED MILKS OF TYPE IV

These are interesting products of very mixed microbial populations. Their study is still in its infancy but since the International Dairy Congress was held in Moscow in 1982 curiosity has been re-kindled.

6.1. Kefir

This is the most famous of the truly alcoholic fermented milks. It is generally reported to have come from the Caucasus mountains and is consumed at a rate of 4·5 kg/person/year in the USSR. Fermentation is started by addition of kefir grains to heat-treated (95°C for 10 min) whole milk. The grains, which resemble cauliflower heads when wet and brownish seeds when dry, are particles of clotted milk plus the kefir organisms. Figure 3 shows a scanning electron micrograph of a grain purchased from a commercial starter manufacturer. Kefir is said to have a complex flora consisting of yeasts and lactic-acid bacteria. Yeasts isolated have been identified as *Saccharomyces delbruekii* (Rosi, 1978a), *Sacc. cerevisiae* (La Riviere, 1963; Rosi, 1978a) and *Sacc. exiguus* (Iwasawa *et al.*, 1982); *Candida* (*Torula*) *kefir* (Kosikowski, 1977) and *Candida pseudotropicalis* (Hirota & Kikuchi, 1976). Of the lactic-acid bacteria *Lb. caucasicus* is the most frequently mentioned. Unfortunately, an authentic strain of this species no longer exists and the epithet *caucasicus* is not recognized because the original culture deposited at the American Type Culture Collection was found to be a mixture of different lactobacilli. Kandler & Kunath (1983) have therefore re-investigated the most frequently isolated heterofermentative lactobacillus and have described it as a new species with the name *Lb. kefir*. These authors (Kunath & Kandler, 1983) examined a number of kefir samples from private households and found that *Lb.*

Fig. 3. Scanning electron micrograph of a kefir grain (×5000).

kefir was the dominant lactobacillus of kefir milk, but that it occurred at only 10% in the kefir grain, the rest of the lactobacillus population at this site being homofermentative *Lb. acidophilus*. Marshall *et al.* (1984a) have also isolated *Lb. kefir* from both 'household' grains and from fresh kefir grains bought from commercial outlets (Hansens Laboratory, Denmark and Valio Co-operative Dairies, Finland), but they found that it was this species which dominated in the grains.

Further studies of the kefir grain (Marshall *et al.*, 1984b) showed that mixed flora of kefir was held together in non-dispersible structures which build up from sheet-like structures (Fig. 4) into large grains (Fig. 5). The extracellular material of the matrix in which the microflora was embedded took up the stains ruthenium red and periodic acid-semicarbazide-silver proteinate, indicating that it was largely composed of fibrillar carbohydrate which is probably branched and composed of glucose and galactose. This fibrillar carbohydrate has been given the name kefiran (La Riviere *et al.*, 1967).

It has been suggested that the carbohydrate is of bacterial origin, produced by a population of lactobacilli which reside within the matrix,

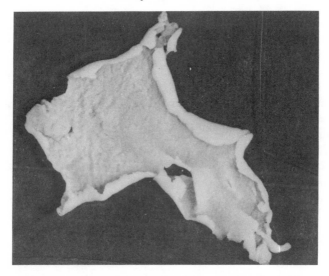

Fig. 4. Sheet-like structure from commercial kefir grains (×6).

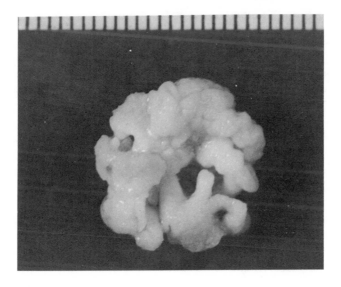

Fig. 5. Large grain from a commercial kefir sample (×5).

and that it separates non-carbohydrate producing populations of lacto-bacilli and yeasts so that sheet-like structures show asymmetry, with yeasts predominating on one side and the obverse almost entirely populated with lactobacilli (Marshall *et al.*, 1984b). Nothing is known of the mechanism of grain formation and attempts at making kefir grains from pure or crude cultures have not been successful (Korolewa, 1975; Hirota & Kikuchi, 1976).

Commercial production of kefir involves using the grains to make the starter culture by inoculating skim milk heated at 95°C for 30 min and incubating at 25°C for 1–2 days. The grains are than recovered for use in making the next starter. The fermented milk is inoculated into whole heat-treated (85°C for 30 min) milk at 5% and fermented for about 12 h at 18–22°C, after which the kefir is ripened for 1–3 days at 10°C before being sold (Kosikowski, 1977).

During the manufacture of kefir milk asepsis is not maintained and differences in microbial composition, e.g. inclusion of species of strep-tococcus, leuconostoc and acetobacter (Rosi, 1978b; Rosi & Rossi, 1978) and coliforms (Vayssier, 1978), may be explained by the diffe-rent techniques employed during processing and/or the different sources of the kefir grain. Vayssier (1978) has catalogued the various bacteria found in kefirs from different countries and calls for a defined starter to be determined. That kefir grains persist under conditions which do not preclude other micro-organisms would suggest that at least part of the microbial composition is specific and constant.

The major end products of the fermentation were reported as lactic acid, ethanol, diacetyl and acetoin, and CO_2 (Kramkowska *et al.*, 1982). A good kefir milk foams like beer and has a smooth pourable consistency (Kosikowski, 1977). Little has been published concerning the inter-relationships between the starter organisms, and the pathways described are conjectural as they are made by inference from know-ledge we have about the individual organisms. For this reason each organism is discussed individually rather than the starter as an entity.

Str. lactis—a variant of this mesophilic starter has been isolated from kefir by Kunath & Kandler (1983). It would be expected to behave in much the same way as it does when used in buttermilk starter (Section 3), using the Embden–Meyerhof pathway, predominantly, for produc-tion of pyruvate. Pyruvate is first convered to lactate and residual pyruvate to diacetyl, and to a lesser extent, acetaldehyde.

Lb. brevis—this was identified by La Riviere *et al.* (1967) and Rosi & Rossi (1978) and is a heterofermentative lactobacillus said to grow only slowly in milk (Bergey, 1974). In mixed culture however, there may be some stimulation. It produces CO_2, acetate and lactate from glucose metabolism via pyruvate and as it is able to utilize citrate may also produce acetoin and diacetyl (reactions 8 and 9, p. 160). These isolates, however, may have been similar to *Lb. kefir* as Vescovo *et al.* (1979) showed that the heterofermentative lactobacilli from their kefir showed high DNA/DNA homology with each other, but not with strains of *Lb. brevis*, a finding confirmed by Kandler & Kunath (1983) and Marshall *et al.* (1984a).

Lb. kefir—this is also a heterofermentative lactobacillus, and when freshly isolated from kefir grains appears to ferment only arabinose, ribose and gluconate (Marshall *et al.*, 1984a). Propagation in laboratory media promoted changes which resulted in loss of ability to ferment arabinose and ability to ferment a wider range of other carbohydrates. Fermentation products included ethanol (up to 0·25%) and CO_2 (Marshall & Cole, unpublished observations).

Lb. acidophilus—this has been isolated as the predominant bacterium of kefir grains by Kunath & Kandler (1983). Its metabolism in milk has been discussed, and its inability to grow well in this medium supports the finding that kefir milk contains *Lb. acidophilus* as the minority species.

Leuconostoc species—likely to be *Leuc. mesenteroides* and *cremoris* and can therefore be expected to behave in similar ways to the buttermilk culture. Keenan & Lindsay (1966) have shown that *Leuc. cremoris* can convert acetaldehyde to ethanol at 5°C. This may contribute to alcohol content when kefir is ripened for 1–3 days at 5–10°C.

Yeasts—lactose fermentation is not essential for kefir fermentation. A yeast isolated by Iwasawa *et al.* (1982) from Danish yeast kefir grains did not ferment lactose and preferentially metabolized galactose in medium containing galactose and glucose. Formation of diacetyl in yeast fermentations has been known for a long time (Suomalainen & Jannes, 1946), and Hommes (1966) has demonstrated acetaldehyde production directly from pyruvate catalysed by pyruvate decarboxylase, an enzyme lacking in the other starter organisms. However, the yeasts also possess an alcohol dehydrogenase to convert acetaldehyde to ethanol (reactions 18 and 19 below) and *Saccharomyces cerevisiae* may combine acetaldehyde with 'active aldehyde' to produce acetoin (reaction 20). Chuang & Collins (1968) have also shown that production of acetoin by yeasts is stimulated by addition of acetaldehyde. The contribution of the yeast population to flavour volatiles in kefir is therefore difficult to assess.

18. pyruvate → acetaldehyde + CO_2

19. acetaldehyde $\xrightarrow{\hspace{2cm}}$ ethanol
$$\overset{NADH \quad NAD}{\underset{}{\smile}}$$

20. acetaldehyde + 'active aldehyde' → acetoin

Yeast–bacterium interactions. Associations between yeasts and lactic acid bacteria are very common in a wide variety of traditional food and beverages and appear remarkably stable, seeming to persist for many years (saké, ginger 'beer', soy sauce). Although little has been published specifically concerning kefir, lactobacillus–yeast interactions have been investigated by the brewing industry, and it is interesting that *Saccharomyces cerevisiae* and *Lb. brevis* aggregate and co-sediment (White & Kidney, 1981). This may have relevance to kefir grain construction, particularly as it has been shown by these authors that Ca^{++} ions promote aggregation.

Whatever the interactions, the end metabolites of associative growth of these organisms result in the final product having a titratable acidity of 0·8% (pH 4·4); acetaldehyde, 1–2 ppm; diacetyl, 3 ppm. Alcohol content is very variable. There are reports of it being as high as 1% (Kosikowski, 1977; Vedamuthu, 1977), but Polish kefir rarely exceeds 0·4% (Pette, 1964) and Finnish and Norwegian kefir have only 0·05% and 0·15% respectively. Reports from Poland and the USSR (Gawel & Gromadka, 1978; Korovinka *et al.*, 1978) give alcohol contents of 0·03% and 0·026%, respectively. Generally the higher the alcohol content, the more yeasty the flavour. Kefir, before ripening at low temperatures, has the aroma of newly baked bread.

6.2. Laban
Baroudi & Collins (1975) have examined the microflora of this fermented milk popular in the Lebanon. *Str. thermophilus, Lb. bulgaricus, Leuc. lactis, Saccharomyces cerevisiae* and *Kluyveromyces fragilis* were identified from a culture that had developed naturally over a number of years. The aroma compound, diacetyl, was not detected organoleptically, and acetaldehyde (4·2 ppm) was attributed not to the *Str. thermophilus*, which produced only small amounts, but to *Kluyveromyces fragilis* (reaction 18 above). *Saccharomyces cerevisiae* produced the alcohol (1·25%). The flavour was described as tart (pH 4·25) and slightly yeasty. These authors also discussed the growth of each

population during fermentation: the milk was inoculated at 50°C with 2·5–3·0% starter so *Str. thermophilus* should start metabolizing first, followed by *Lb. acidophilus, Leuc. lactis* and finally the yeasts as the temperature cooled to 20–22°C.

6.3. Koumiss

Traditionally this was made from mares' milk but because of its popularity the USSR is now making it from cows' milk which has been fortified with whey protein. This fortification prevents syneresis and greater alcohol and CO_2 production by starter is achieved. It is fermented with a mixed starter of *Lb. bulgaricus* and *Candida* (*Torula*) *kefir* (Kosikowski, 1977) and may contain up to 3% alcohol. A further refinement is to distil koumiss into a highly intoxicating beverage.

Koumiss prepared from unfortified cows' milk in our laboratory using a starter consisting of *Saccharomyces cerevisiae* and *Lb. bulgaricus* had a pleasant flavour and aroma. It had a pH 4·1, acetaldehyde at 50 ppm and 0·4% ethanol when formate was added to the milk.

7. OTHER FLAVOURS AND DEFECTS

Differences in the milk of each species are reflected in the composition and therefore the flavour of its milk. Mares' milk is high in lactose (6·2%) and has a greater proportion of whey protein compared with cows' milk. Ewes' and buffalo milks have a high percentage of fat. Fat content makes a large difference to both texture and flavour and to the tendency to produce rancidity, so it matters whether the product is made with full-fat or skim milk. Milk from mastitic quarters or from late lactation cows may have a salted flavour (Rasic & Kurmann, 1978). This may, however, be diluted out when milk is mixed with milk from other cows.

7.1. Off-flavours

It is worth noting that flavour differences due to the presence of highly volatile compounds in milk before processing may not affect the final product because a high-heat treatment (80–95°C for 10–30 min) often precedes fermentation and if given in an open system will drive off these volatiles.

This, however, may produce a 'cooked' flavour or result in caramelization of sugars. Release of peptides and glycopeptides also occurs, their release being dependent on the temperature and duration of the heat treatment (Hindle & Wheelock, 1970).

The influence of feed on the appearance of a number of taints has been investigated and some interesting flavours have been reported, ranging from raspberry to blackcurrant. These are a consequence of ketone formation in the rumen, resulting from a shift in microbial metabolism. These and other flavour volatiles (mostly ketones) have been discussed by Forss (1978).

Off-flavours can be derived from protein, fat or sugar degradation. Proteolysis often leads to the production of bitter peptides and adjectives such as 'unclean' have often been used to describe the defect. Further degradation of the peptides, however, may remove bitterness but it may be replaced by other undesirable taints. A putrid flavour in cultured buttermilk, for example, results from decomposition of the amino acid, tryptophan to skatole and indole. These compounds contribute to the butter flavour (Forss *et al.*, 1967), but in excess lead to offensive defects (Kosikowski, 1977). Increased proteolysis may occur if milks are contaminated with psychrotrophs (Cousin & Marth, 1977) as many of the proteases resist the heat-treatment given to the milk before starter inoculation, and most lactic-acid bacteria (i.e. streptococci, lactobacilli and leuconostocs) possess lipase activity (Fryer *et al.*, 1967). Fat breakdown gives rise to rancidity to a greater or lesser degree, and high levels of free fatty acids may also affect starter activity. Capric acid has been found to have an inhibitory effect on proteolysis by *Lb. bulgaricus* (Poznanski *et al.*, 1968), whereas oleic acid has a marked effect on pyruvate metabolism: oleic acid prevents acetate formation in the group N streptococci and at low pH acetoin and diacetyl are produced (Anders & Jago, 1970). Fatty acids may also be esterified and produce fruity flavours. Bills *et al.* (1965) identified the compounds responsible for the fruity defect in Cheddar cheese as ethyl butyrate and ethyl hexanoate. These authors correlated high ester formation with high ethanol content. Esterases accelerate the formation of these compounds: Hosano *et al.* (1974) have shown that 5 strains of lactic-acid bacteria and 2 psychrotrophic bacteria exhibit such activity. Thus starter metabolism may be responsible for poor quality.

8. MODERN TECHNOLOGY AND THE FUTURE OF FERMENTED MILKS

In countries where fermentation is not a tradition, the acceptance of fermented milks has been slow. Yoghurt is consumed mainly as a snack or dessert in Western Europe and North America. The renewed interest in kefir has led to the manufacture of this product in greater quantities in Norway and Sweden and also to its manufacture in West Germany. The USSR is also making koumiss from cows' milk, in addition to mares milk, in response to increased consumer demand. Carbonated yoghurt has been developed which has a thirst-quenching character similar to kefir (Duitschaever & Ketcheson, 1974). These countries use pure culturing methods, mistrusting natural souring or the 'back-slop' procedure.

8.1. Modern Manufacture

The consumer demands from the manufacturer a product with consistent flavour and texture; technology has answered these demands by providing good starters and making the product in large technologically advanced plants where temperatures are accurately controlled. The post World War II development of a wide variety of yoghurts has been remarkable. Mann (1981) has compiled a digest of recent publications covering developments in yoghurt manufacture. Powdered 'instant' yoghurt is now available covered by a British patent (Yoshimi, 1979). Other technological achievements include a US patent for the manufacture of yoghurt by direct acidification (Igoe, 1979) and making yoghurt by continuous culture. This can be based on continuous incubation and cooling (Hansen, 1977) or by a two stage process where growth of the starter is limited to a pH of 5·7 by injection of fresh milk. The second stage involves further acidification and formation of texture (MacBean *et al.*, 1978, 1979). A closer study of this method has resulted in an apparatus for continuous manufacture of yoghurt at 4000 litres/h being assembled at NIZO, The Netherlands (Driessen *et al.*, 1982). Improvements to starter vitality can still be made. There has been a move towards the preparation of concentrates for direct inoculation of the process milk, i.e. the concentrate = bulk starter. Only small losses in viability and cell activity can be tolerated when using this method. The finding that starter proteolytic enzymes may be reduced when freezing at $-196°C$ is used to prepare these concentrates (Petterssen, 1975) must have relevance in the preparation of yoghurt starter which relies on good proteolytic activity for controlled mixed population growth.

Recently a single starter yoghurt consisting of *Lb. bulgaricus* has been used to ferment milk containing 17% total solids to which formate has been added to stimulate growth (Marshall & Mabbitt,

1980). Fermentation was carried out at 42°C for 4–6 h and resulted in a product with a yoghurt-like consistency and aroma (22–35 ppm acetaldehyde) and sharp taste (pH 4·2–4·4). Taste panel assessment of this product suggested that there was little diacetyl present (buttery flavour and aroma) and no ethanol was demonstrated on gas chromatographic analysis (Marshall *et al.*, 1982). There is a report, however, that *Lb. bulgaricus* can produce up to 13 ppm diacetyl (Dutta *et al.*, 1973).

Yoghurt made from ultrafiltered milk makes a finer, smoother product (Bundgaard *et al.*, 1972; Chapman *et al.*, 1974) as does milk concentrated by reverse osmosis (Davies *et al.*, 1977). Ultrafiltration has also been suggested for the manufacture of koumiss (Puhan & Gallmann, 1980). Further advance, however, can only come from a more rigorous approach. To quote Kosikowski (1977) 'The aromatic acid flavours of a good yoghurt or cultured buttermilk are as subtle as French perfume'. These subtleties deserve investigation as it is these differences which keep interest alive in these valuable foods.

8.2. Biochemical Investigations on Starter Organisms

The mathematical approach of Driessen and his colleagues (1982) has told us more about the inter-relationships between *Str. thermophilus* and *Lb. bulgaricus*. Their work has been directed towards more efficient production of yoghurt by continuous flow, but flavour development should not be neglected and the relationships which lead to flavour production in mixed cultures require further study.

The metabolic products of a given population in a mixed starter may not only promote growth but under certain circumstances may limit it and promote flavour production. Growth limitation may also be desirable in a mixed starter containing many types (such as kefir and laban) to prevent domination by one population. Bergère & Hermier (1968) noted that sensitivity of *Str. cremoris* to lactate and hydrogen peroxide is inhibitory to many bacteria (Gilliland & Speck, 1969; 1972). Moon & Reinbold (1976) have demonstrated an inhibitory effect on *Lb. bulgaricus* exerted by *Str. thermophilus* in stationary phase. Alcohol is toxic, as is acetaldehyde, both are quickly eliminated by most bacteria and these will also limit growth. Organisms which have different tolerances to these compounds may be worth investigation. The control of metabolic pathways in *Str. lactis* has been studied by Thompson & Thomas (1977) and Thompson (1978) and further investigations may lead to possibilities of controlling flavour production and mixed starter growth.

It is possible that manipulation of metabolic pathways may prevent spoilage. H_2O_2 is produced by *Lb. lactis* (Wheater *et al.*, 1952), nisin is produced by the streptococci, bulgarican or acidophilin by the lactobacilli (Shahani *et al.*, 1976, 1977) and lactacin B by *Lb. acidophilus* (Barefoot & Klaenhammer, 1983). The presence of such substances may obviate the need for pasteurizing the final product. The high alcohol content of some products (laban, koumiss) may be turned to advantage in the production of fruity esters if the microbial flora contains sufficient esterase activity. Esterase and alcohol dehydrogenases may be gained or lost, stimulated or inhibited relatively simply as they are at the end of secondary metabolic pathways. These approaches are exciting paths to follow towards the development of new and improved products.

The future of fermented milks and their cultures lies therefore in genetic studies and in control of metabolic pathways to improve flavour production and strain vitality. There is no doubt that food shortages on a global basis will persist and investment in fermented foods is necessary.

9. CONCLUSIONS

Cultured dairy products are accepted by the consumer mainly on the basis of flavour. During the last twenty years considerable advances have been made towards an understanding of cultured product flavours and the role of micro-organisms in the formation and degradation of these flavours.

The development of quantitative gas chromatographic analysis of cultures has resulted in a more exact evaluation of flavour volatiles (Bills *et al.*, 1965; Lindsay *et al.*, 1965) so that formulation of mixed starters can be based on their ability to produce diacetyl or acetaldehyde as well as on their ability to produce lactic acid. Knowledge of fermentation pathways leading to the flavour volatiles permits manipulation of these routes. For example, addition of a single amino acid, threonine, increases acetaldehyde production in strains of *Lb. acidophilus* and bidifobacteria, and manufacture of a fermented milk with a yoghurt flavour is now possible using these organisms (Marshall *et al.*, 1982; Marshall & Cole, 1983).

The importance of mixed cultures in the dairy industry is well established but the stability of starters for yoghurt and kefir are only just being appreciated. Models for the interaction of *Str. thermophilus* and *Lb. bulgaricus* have been formulated (Meyer *et al.*, 1975; Ubbels, 1981) but studies involving yeast/bacterium interactions are still in their infancy. While consumption of fermented milks continues to increase there is good incentive to expand the range and quality of fermented products through studies of flavour volatiles, growth and interaction of many useful micro-organisms.

REFERENCES

ABRAHAMSEN, R. K. (1978) *XX Int. Dairy Congress* E 829.

ACCOLAS, J. P., BLOQUEL, R., DIDIENNE, R. & REGNIER, J. (1977). *Le Lait* **57**, 1.

ALM, L. (1981) *J. Dairy Sci. Food Agric.* **32**, 1247.

ANDERS, R. F. & JAGO, G. R. (1970) *J. Dairy Res.* **37**, 445.

AYEBO, A. D., ANGELO, I. A. & SHAHANI, K. M. (1980) *Milchwissenschaft* **35**, 730.

BAREFOOT, S. F. & KLAENHAMMER, T. R. (1983) *Appl. Environ. Microbiol.* **45**, 1808.

BAROUDI, A. A. G. & COLLINS, E. B. (1975) *J. Dairy Sci.* **59**, 200.

BERGÈRE, J.-L. & HERMIER, J. (1968) *Le Lait* **48**, 13.

BERGEY, D. H. (1974) *Bergey's Manual of Determinative Bacteriology*, R. E. Buchanan & N. E. Gibbons, Eds, 8th edn. Williams & Wilkins Co., Baltimore.

BERRIDGE, N. J. (1966) *J. Soc. Dairy Technol.* **19**, 232.

BERTELSOL, E. & JONSSON, U. (1976) *Svenska Mejeritidningen* **68**, 9.

BHATTACHARYA, D. C., RAJ, D. & TIWARI, B. D. (1980) *Indian J. Dairy Sci.* **33**, 38.

BILLS, D. D., MORGAN, M. E., LIBBEY, L. M. & DAY, E. A. (1965) *J. Dairy Sci.* **48**, 765.

BOTTAZZI, V. & VESCOVO, M. (1969) *Netherlands Milk Dairy J.* **23**, 71.

BOTTAZZI, V., BATTISTOTTI, B. & MONTESCANI, G. (1973) *Le Lait* **53**, 295.

BUNDGAARD, A. G., OLSEN, O. J. & MADSEN, R. F. (1972) *Dairy Industries* **37**, 539 & 544.

CARDENAS, I., DE RUIZ HOLGARDO, A. P. and OLIVER, G. (1980) *Milchwissenschaft* **35**, 296.

CHAPMAN, H. R., BINES, V. E., GLOVER, F. A. & SKUDDER, P. J. (1974) *J. Soc. Dairy Technol.* **27**, 151.

CHUANG, L. F. & COLLINS, E. B. (1968) *J. Bacteriol.* **95**, 2083.

COLLINS, E. B. (1972) *J. Dairy Sci.* **55**, 1022.

COLLINS, E. B. & BRUHN, J. C. (1970) *J. Bacteriol.* **103**, 541.

CONN, H. W. (1900) *Annual Report of the Storrs School*, cited by Sandine *et al.* (1972) *J. Milk Food Technol.* **35**, 176.

COUSIN, M. A. & MARTH, E. H. (1977) *Cultured Dairy Products J.* **12**, 30.

DAVIES, F. L., SHANKAR, P. A. & UNDERWOOD, H. M. (1977) *J. Soc. Dairy Technol.* **30**, 23.

DEETH, H. C. & TAMIME, A. Y. (1981) *J. Food Protection* **44**, 78.
DELANEY, A-G. O. (1977) *Milchwissenschaft* **32**, 651.
DE MANN, J. M. & GALESLOOT, T. E. (1962) *Netherlands Milk Dairy J.* **16**, 1.
DRIESSEN, F. M. (1981) In *Mixed Culture Fermentations*, D. Bushell & L. Slater, Eds. Academic Press, London.
DRIESSEN, F. M., KINGMA, F. & STADHOUDERS, J. (1982) *Netherlands Milk Dairy J.* **36**, 135.
DUITSCHAEVER, C. L. (1978) *Cultured Dairy Products J.* **13**, 20.
DUITSCHAEVER, C. L. & KETCHESON, G. (1974) *Dairy Ice-Cream Field* **157**, 66H.
DUTTA, S. M., KUILA, R. K. & RANGANATHAN, B. (1973) *Milchwissenschaft* **28**, 231.
FORSS, D. A. (1978) *XX Int. Dairy Congress*, 78ST.
FORSS, D. A., STARK, W. & URBACH, G. (1967) *J. Dairy Res.* **34**, 131.
FOWLER, G. G. (1969) *Milchwissenschaft* **24**, 211.
FRYER, T. F., REITER, B. & LAWRENCE, R. C. (1967) *J. Dairy Sci.* **50**, 388.
GALESLOOT, TH. E., HASSING, F. & VERINGA, H. A. (1968) *Netherlands Milk Diary J.* **22**, 50.
GARVIE, E. I. (1983) *Int. J. Systematic Bacteriol.* **33**, 118.
GAWEL, J. & GROMADKA, M. (1978) *XX Int. Dairy Congress* E 839.
GILLILAND, S. E. & SPECK, M. L. (1969) *Appl. Microbiol.* **17**, 797.
GILLILAND, S. E. & SPECK, M. L. (1972) *J. Milk Food Technol.* **35**, 307.
GILLILAND, S. E. & SPECK, M. L. (1977) *J. Dairy Sci.* **60**, 1394.
GILLILAND, S. E., ANNA, E. D. & SPECK, M. L. (1970) *Appl. Microbiol.* **19**, 890.
GILLILAND, S. E., SPECK, M. L., NAUYOK, G. F. & GIESBRECHT, F. G. (1978) *J. Dairy Sci.* **61**, 1.
GORNER, F., PALO, V. & BERTAN, M. (1968) *Milchwissenschaft* **23**, 94.
GROUX, M. (1973) *Le Lait* **53**, 146.
GYLLENBERG, H. & ROINE, P. (1957) *Acta Pathologica Microbiologica Scandinavica* **41**, 144.
HAENEL, H. (1970) *American J. Clinical Nutrition* **23**, 1433.
HAMDAN, I. Y., KUNSMAN, J. E. & DEANE, D. D. (1971) *J. Dairy Sci.* **54**, 1080.
HANSEN, R. (1977) *Nordeuropaeisk Mejeri Tidsskrift* **43**, 274.
HARVEY, R. J. & COLLINS, E. B. (1963) *J. Biol. Chem.* **238**, 2648.
HEGAZI, F. Z. & ABO-ELNAGA, I. G. (1980) *Z. Lebensmitteluntersuchung und Forschung* **171**, 367.
HINDLE, E. J. & WHEELOCK, J. V. (1970) *J. Dairy Res.* **37**, 397.
HIROTA, T. & KIKUCHI, T. (1976) Reports of research of Snow Brand Milk Products Co. Laboratory No. 74, 63.
HOMMES, F. A. (1966) *Arch. Biochem. Biophys.* **114**, 231.
HOSANO, A., ELLIOTT, J. A. & MCGUGAN, W. A. (1974) *J. Dairy Sci.* **57**, 535.
HUMPHRIES, C. L. & PLUNKETT, M. (1969) *Dairy Sci. Abstr.* **31**, 607.
IGOE, R. S. (1979) US Patent 4 169 854.
INTERNATIONAL DAIRY FEDERATION (1983) *Cultured dairy foods in human nutrition*, Document 159.

IWASAWA, S., UEDA, M., MIYATA, N., HIROTA, T. & AHIKO, K. (1982) *Agric. Biol. Chem.* **46**, 2631.

KANDLER, O. & KUNATH, P. (1983) *Systematic Appl. Microbiol.* **4**, 286.

KEENAN, T. W. & BILLS, D. D. (1968) *J. Dairy Sci.* **51**, 1561.

KEENAN, T. W. & LINDSAY, R. C. (1966) *J. Dairy Sci.* **50**, 1585.

KEENAN, T. W., LINDSAY, R. C., MORGAN, M. E. & DAY, E. A. (1966) *J. Dairy Sci.* **49**, 10.

KHANA, A. & SINGH, J. (1979) *J. Dairy Res.* **46**, 681.

KOROLEWA, N. S. (1975) cited by Molska *et al.* (1980) *Acta Alimentaria Polonica* **6**, 145.

KOROVINKA, L. N., PATKUL, G. M. & MASLOV, A. M. (1978) *XX Int. Dairy Congress* 841.

KOSIKOWSKI, F. V. (1977) *Cheese and Fermented Milk Foods,* 2nd edn. Edwards Brothers, Michigan.

KOSIKOWSKI, F. V. (1978) *Cultured Dairy Products J.* **13**, 5.

KRAMKOWSKA, A., KORNACKI, K., BAUMAN, B. & FESNAK, D. (1982) *XXI Int. Dairy Congress* **1**, 304.

KUILA, R. K. & RANGANATHAN, B. (1978) *J. Dairy Sci.* **61**, 379.

KUNATH, P. and KANDLER, O. (1983) Lactic acid bacteria in foods. Symposium of the Netherlands Society for Microbiology, Wageningen, The Netherlands.

LA RIVIERE, J. W. M. (1963) *J. General Microbiol.* **31**, (v).

LA RIVIERE, J. W. M., KOOIMAN, P. & SCHMIDT, K. (1967) *Archiv fur Microbiologie* **59**, 269.

LANG, F. (1980) *Milk Industries* **82**, 44.

LEES, G. J. & JAGO, G. R. (1976) *J. Dairy Res.* **43**, 75.

LEES, G. J. & JAGO, G. R. (1977) *J. Dairy Res.* **44**, 139.

LEES, G. J. & JAGO, G. R. (1978a) *J. Dairy Sci.* **61**, 1205.

LEES, G. J. & JAGO, G. R. (1978b) *J. Dairy Sci.* **61**, 1216.

LINDSAY, R. C., DAY, E. A. & SANDINE, W. E. (1965) *J. Dairy Sci.* **48**, 863.

LINDSAY, R. C., DAY, E. A. & SATHER, L. A. (1966) *J. Dairy Sci.* **50**, 25.

LODDER, J. (1970) *The Yeast,* J. Lodder, Ed. North Holland Publishing Co., Amsterdam.

MACBEAN, R. D., LINKLATER, P. M. & HALL, R. J. (1978) *XX Int. Dairy Congress* E 827.

MACBEAN, R. D., LINKLATER, P. M. & HALL, R. J. (1979) *Biotechnol. Bioengineer.* **21**, 1517.

MADA, M. (1981) *Japanese J. Dairy Food Sci.* **30**, A205.

MADA, M. (1982) *Japanese J. Dairy Food Sci.* **31**, A253.

MAHLER, H. R. & CORDES, E. H. (1971) *Biological Chemistry,* 2nd edn. Harper & Row, London and New York, pp. 117–341.

MANN, E. J. (1981) *Dairy Industries Int.* **46**, 28.

MANUS, L. J. (1979) *Cultured Dairy Products J.* **14**, 9.

MARIER, J. R. & BOULET, M. (1958) *J. Dairy Sci.* **41**, 1683.

MARSHALL, V. M. & COLE, W. M. (1983) *J. Dairy Res.* **50**, 375.

MARSHALL, V. M. & MABBITT, L. A. (1980) *J. Soc. Dairy Technol.* **33**, 129.

MARSHALL, V. M. E., COLE, W. M. & VEGA, J. R. (1982) *J. Dairy Res.* **49**, 665.

MARSHALL, V. M., COLE, W. M. & FARROW, J. A. E. (1984a) *J. Appl. Bacteriol.* **56**, 503.
MARSHALL, V. M., COLE, W. M. & BROOKER, B. E. (1984b) *J. Appl. Bacteriol.* (In press).
MCKAY, L. & BALDWIN, K. A. (1974) *J. Dairy Sci.* **57**, 181.
MEYER, J. S., TSUCHIYA, H. M. & FREDRICKSON, A. G. (1975) *Biotechnol. Bioengineer.* **17**, 1065.
MILLER, I. & KANDLER, O. (1964) *Medizin und Ernahrung* **5**, 100.
MILLER, I., MARTIN, H. & KANDLER, O. (1964) *Milchwissenschaft* **19**, 18.
MOON, N. J. & REINBOLD, G. W. (1976) *J. Milk Food Technol.* **39**, 337.
NILSSON, R. & NILSSON, G. (1958) *Archiv fur Mikrobiologie* **31**, 191.
ODUM, E. P. (1971) *Fundamentals of Ecology*, W. B. Saunders, Ed. Philadelphia, London and Toronto, p. 211.
O'LEARY, V. S. & WOYCHICK, J. H. (1976) *Appl. Environ. Microbiol.* **32**, 89.
ORLA–JENSEN, S. (1931) *Dairy Bacteriology*, 2nd edn., J. & A. Churchill, London.
PETTE, J. W. (1964) Bulletin FIL/IDF, part III, report no. 7.
PETTE, J. W. & LOLKEMA, H. (1950a) *Netherlands Milk Dairy J.* **4**, 197.
PETTE, J. W. & LOLKEMA, H. (1950b) *Netherlands Milk Dairy J.* **4**, 209.
PETTE, J. W. & LOLKEMA, H. (1950c) *Netherlands Milk Dairy J.* **4**, 261.
PETTERSSEN, H. E. (1975) *Appl. Microbiol.* **29**, 133.
POZNANSKI, S., SURAZNSKI, A. & OBYRN, T. (1968) *Le Lait* **48**, 261.
PREMI, L., SANDINE, W. E. & ELLIKER, P. R. (1972) *Appl. Microbiol.* **24**, 51.
PUHAN, Z. & GALLMANN, P. (1980) *Cultured Dairy Products J.* **15**, 12.
RASIC, J. L. & KURMANN, J. A. (1978) *Yoghurt*. Published by the authors, Technical Dairy Publishing House, Copenhagen.
ROSI, J. (1978a) *Scienza e Technica Lattiero-Casearia* **29**, 59.
ROSI, J. (1978b) *Scienza e Technica Lattiero-Casearia* **29**, 221.
ROSI, J. & ROSSI, J. (1978) *Scienza e Technica Lattiero-Casearia* **29**, 291.
SCHULTZ, M. E. & HINGST, G. (1954) *Milchwissenschaft* **9**, 330.
SHAHANI, K. M., VAKIL, J. R. & KILARA, A. (1976) *Cultured Dairy Products J.* **11**, 14.
SHAHANI, K. M., VAKIL, J. R. & KILARA, A. (1977) *Cultured Dairy Products J.* **12**, 8.
SHANKAR, P. (1977) PhD Thesis, University of Reading.
SHANKAR, P. A. & DAVIES, F. L. (1977) *J. Soc. Dairy Technol.* **30**, 28.
SOUMALAINEN, H. & JANNES, L. (1946) *Nature* **157**, 336.
SPECK, M. L. (1975) *J. Dairy Sci.* **58**, 783.
SPECK, M. L. (1976) *J. Dairy Sci.* **59**, 338.
SPECKMAN, R. A. & COLLINS, E. B. (1968) *J. Bacteriol.* **95**, 174.
SUNDMAN, V. (1953) *Int. Dairy Congress* **3**, 1420.
TAMIME, A. Y. & DEETH, H. C. (1980) *J. Food Protection* **43**, 939.
TINSON, W., HILLIER, A. J. & JAGO, G. R. (1982) *Australian J. Dairy Technol.* **37**, 8.
THOMPSON, J. (1978) *J. Bacteriol.* **136**, 465.
THOMPSON, J. & THOMAS, T. D. (1977) *J. Bacteriol.* **130**, 583.
UBBELS, J. (1981) Cited by Driessen F. M. (1981) in *Mixed Culture Fermentations*, D. Bushell & L. Slater, Eds. Academic Press, London.

VAYSSIER, Y. (1978) *Revue Laitiere Francais* **361,** 73.
VEDAMUTHU, E. R. (1977) *J. Food Protection* **40,** 801.
VERINGA, H. A., GALESLOOT, TH. E. & DAVELAAR, H. (1968) *Netherlands Milk Dairy J.* **22,** 114.
VESCOVO, M., DELLAGLIO, F., BOTTAZZI, V. & SARRA, P. G. (1979) *Microbiologica* **2,** 317.
WHEATER, D. M., HIRSCH, A. & MATTICK, A. T. R. (1952) *Nature* **170,** 623.
WHITE, F. H. & KIDNEY, E. (1981) *Mixed Culture Fermentations,* M. E. Bushell & J. H. Slater, Eds. Academic Press, London.
YOSHIMI, T. O. (1979) UK Patent 2 018 124A.

Chapter 7

Flavour Development in Cheeses

BARRY A. LAW

National Institute for Research in Dairying, Shinfield, Reading, UK

1. INTRODUCTION

The mechanisms by which relatively tasteless cheese curds become distinctively flavoured cheeses have been under research scrutiny for many decades. Despite all this attention, the cheeses whose maturation processes are well understood are the exception rather than the rule. It is generally appreciated, however, that the course of cheese maturation is governed by an interplay of several different factors including pH and moisture levels reached during the initial conversion of milk to curds, the level and method of salt addition, the ripening temperature and the nature of the secondary (non-starter) microflora which develops in or on the cheese. Although the similarities and differences between cheese varieties have been reviewed recently (Law, 1982), a brief summary of the distinguishing features of the main categories is included here for the sake of completeness (Table 1).

High moisture cheeses (50–80%) are termed 'soft'; some are consumed fresh (e.g. cottage cheese, Mozzarella) while others develop a surface flora of moulds and yeasts (e.g. Brie, Camembert). Semi-hard and semi-soft cheeses have moistures in the range 45–50% and include such varieties as mild-flavoured Caerphilly and Edam, with relatively simple lactic floras, contrasting with Limburger and Munster with complex surface floras and very strong flavours. Blue vein cheeses are in the semi-hard category, being distinguished by the growth of *Penicillium roqueforti* within the cheese matrix during maturation. These are not chance contaminants, but are specially selected pure strains, added as spores to the cheese milk. The hard cheeses are made by processes

TABLE 1
Major Cheese Categories, their Starter Compositions and Secondary Microfloras

Cheese category	Example varieties	Moisture content (%)	Starter composition	Starter function	Secondary flora	Major flavour compounds
Unripened, soft	Cottage	not >80	Streptococcus diacetylactis, Leuconostoc spp.	Acid and diacetyl production	None	Lactic acid, diacetyl, acetaldehyde
	Mozzarella	>50	Str. thermophilus, Lactobacillus bulgaricus	Acid production	None	Lactic acid
Ripened, soft	Camembert Brie	48 55	Str. cremoris, Str. lactis	Acid production	Penicillium caseicolum, yeasts	Fatty acids, ammonia, aromatic hydrocarbons, oct-1-en-3-ol, bis (methyl thiomethane), phenylethanol
Semi-soft	Caerphilly	45	Str. diacetylactis, Leuconostoc, Str. cremoris, Str. lactis	Acid and diacetyl production	Lactobacilli	Lactic acid, diacetyl
	Limburg	45	Str. lactis, Str. cremoris	Acid production	Yeasts, Brevibacterium linens	Amino acids, fatty acids, ammonia, methanethiol, acetyl methyl disulphide
Semi-hard	Gouda	40	Str. cremoris, Str. lactis, Str. diacetylactis, Leuconostoc	Acid and carbon dioxide production	Propionibacteria[a]	Amino acids, fatty acids

Type	Cheese	Temp. (°C)	Starter organisms	Function	Associated organisms[a]	Products/characteristics
Blue-vein	Roquefort, Gorgonzola, Stilton, Danish blue	40–45	*Str. lactis* *Str. diacetylactis* *Str. cremoris* *Leuconostoc* spp.	Acid and carbon dioxide production	*Penicillium roqueforti,* yeasts, micrococci	Fatty acids, ketones, lactones, aromatic hydrocarbons
Hard	Cheddar	<40	*Str. cremoris* *Str. lactis* *Str. diacetylactis*[b] *Leuconostoc* spp.	Acid production	Lactobacilli, pediococci	Amino acids, fatty acids, alcohols, pentanone, hydrogen sulphide, methanethiol
	Emmental	38	*Str. thermophilus* *Lb. helveticus* *Lb. lactis* *Lb. bulgaricus* *Propionibacterium shermanii*	Acid, carbon dioxide and propionic acid production	*Propionibacterium shermanii,* group D streptococci	Amino acids (especially proline), peptides, butyric acid, acetic acid, methanethiol thioesters, dimethyl sulphide, alkyl pyrazines
	Gruyère	38–40	*Str. thermophilus* *Lb. helveticus* *Lb. lactis* *Lb. bulgaricus* *Propionibacterium shermanii*[c]	Acid, carbon dioxide and propionic acid production	*Propionibacterium shermanni,* group D streptococci plus yeasts and coryneforms including *Br. linens*	

[a] Sometimes present as adventitious bacteria but not vital to typical cheese characteristics (Kleter, 1976); [b] not always included; [c] introduced with the starter but have no lactic acid-forming function (they grow as a secondary flora).

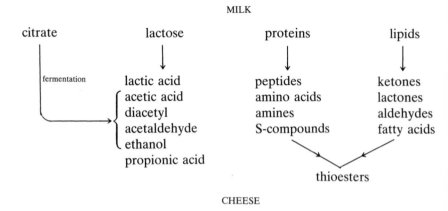

Fig. 1. Major flavour forming pathways in cheese manufacture

which include a heating ('scalding') stage applied to the curds. This may be up to 40°C for cheeses made with mesophilic starters (*Streptococcus cremoris* and *lactis*) or 50°C for those made with thermophilic cultures (*Str. thermophilus* and lactobacilli; see Table 1). Secondary floras range from non-starter lactic-acid bacteria (lactobacilli and pediococci) to propionibacteria, surface yeasts and brevibacteria.

The present discussion will attempt to draw together present knowledge and speculation concerning the conversions which the constituents of milk undergo in the course of cheese ripening, and the possible role of their breakdown products in cheese flavour profiles. The major degradative pathways to be considered are summarized in Fig. 1.

2. INFLUENCE OF CARBOHYDRATES ON FLAVOUR PRODUCTION

2.1. Lactose

Pathways of lactose utilization and metabolism to lactic acid have been dealt with at length in Chapter 3. Their significance in the development of flavour in unripened cheeses such as cottage cheese is limited to the production of the sharp lactic acid flavour. Production of aroma compounds is normally associated with the metabolism of citrate.

Lactose fermentation in ripened cheeses has less obvious but important effects on the course of their maturation. Lactic acid has a stabilizing effect on cheese by virtue of its antibacterial properties (Babel, 1977) and its effect in lowering the cheese redox potential and pH. This ensures that enzymic reactions proceed only slowly, and although most cheeses need long storage times to develop full flavour, they have the desirable property of remaining palatable for many months. Low redox potential also ensures that flavourful sulphur compounds remain in their reduced form.

The fermentation of residual lactose in freshly-pressed cheese curd by secondary lactic-acid bacteria (lactobacilli, pediococci) has been directly related to the quality of Cheddar cheese (Fryer, 1982). These organisms tend to metabolize lactose by heterofermentative pathways under the suboptimal growth conditions in cheese, over-producing compounds with the potential to impair flavour balance and give rise to defects (e.g. formic acid, ethanol, acetic acid; Thomas *et al.* 1979).

Cheese with clean, uniform quality is more likely to result from manufacturing conditions which allow the starter streptococci to use up virtually all the lactose. Turner & Thomas (1980) showed that this can be done by careful salting of the curd so that the salt-in-moisture (SM) level is as close as possible to 4%; the starter bacteria can then go on metabolizing the lactose to lactic acid within a further 24 h in the pressed curd. At high SM levels, starter metabolism is inhibited and lactose persists to provide a substrate for secondary floras.

An alternative approach suggested by Fryer (1982) involves the rapid cooling of freshly pressed curds to 10°C so that secondary lactic-acid bacteria grow only slowly and reach populations of $10^6/g$ or less in the cheese. Under these conditions they metabolize residual lactose homofermentatively to lactic acid and do not produce the compounds which could impair flavour balance. The precise concentrations of these products which must be generated in order to cause defects are not known for Cheddar cheese but, ironically, their complete elimination is also detrimental to desirable flavour (Czulak *et al.*, 1974). In practice a compromise between extremes of over- or under-protection is hoped for and achieved by accident rather than by design.

In contrast to Cheddar cheese, the fermentation of lactose to lactic acid in Emmental and Gruyère cheese by *Str. thermophilus* and *Lactobacillus helveticus* actually provides the substrate for the flavour-producing secondary flora of deliberately added *Propionibacterium shermanii*; its growth and metabolism have to be considered together because they are interdependent. The *Str. thermophilus/Lb. helveticus* starter grows and produces lactic acid both in the cheese vat and during pressing, but the most important stage is recognized as the 24 h draining and pressing time during which the cheese cools from 50°C to 20°C. Accolas *et al.* (1978) and Moquot (1979) have described the effects of the temperature gradient in cooling Emmental cheese on the acid production, first by *Str. thermophilus*, then by *Lb. helveticus* such that after 24 h all the lactose is fermented to lactic acid, more being produced near the cool periphery than in the warmer centre of the cheese. This lactate gradient, together with the salt gradient which results from brining the young cheese, ensures that the propionibacteria grow best near the centre since they are inhibited by high salt and high lactate concentrations. The rate of growth and CO_2 production of this secondary flora is very critical because it governs the size and distribution of holes ('eyes') in the cheese and hence its market value. Emmental should have larger holes than Gruyère, an objective reached by using higher ripening temperatures and lower salting rates for the former variety (Moquot, 1979).

Fermentation of lactate and residual sugars by propionic-acid bacteria is thus a vital stage in Swiss-cheese ripening, and follows initial lactic fermentation by the starters. Hettinga & Reinbold (1972) have reviewed the extensive literature on the propionate fermentation, which involves a complex double cycle via pyruvate, catalytic amounts of methylmalonyl-CoA, and an apparently unique biotin-dependent transcarboxylation reaction. The products are carbon dioxide (which forms the 'eyes'), and propionic and acetic acids, which, with lactic acid, dominate the flavour of Emmental cheese in particular. Eye formation begins after the propionibacteria reach maximum population (from 25–40 days and onwards) while the texture of the cheese is sufficiently plastic to deform, rather than crack. So called 'secondary fermentation' causes excessive CO_2 formation in older cheeses, and leads to such texture defects. It appears to be caused by late stimulation of propionibacteria by amino acids released by the proteolytic action of certain starter cultures or by undesirable contaminants (Moquot, 1979; Steffen, 1979).

The development of surface floras on soft and semi-soft cheeses is dependent on lactose metabolism to lactic acid since it is thought to be initiated by lactate-utilizing yeasts whose growth raises the pH and allows other organisms to colonize the surface. Such a sequence of events is universally recognized for surface smear cheeses such as

Limburg, Tellagio and Gruyère cheese. Yeast typical of these cheese floras include species of *Geotrichium* and *Candida*. On Tellagio cheese both micrococci and coagulase-negative staphylococci form the major pigmented flora (Todesco *et al.*, 1981) but *Brevibacterium linens* is important on Limburg and Gruyère (Purko *et al.*, 1951; Accolas *et al.*, 1980; Boyoval & Desmazeaud, 1983). The role of such organisms in flavour development of cheeses will be discussed in Sections 3 and 4.

2.2. Citrate

Citrate is a normal, if minor component of milk and can be metabolized by *Str. lactis* var. *diacetylactis* and leuconostocs to CO_2 and carbonyl compounds including diacetyl (see Chapter 3). Strains of these organisms are therefore always present in starter cultures for cottage cheese. The flavour of this variety is dependent chiefly on diacetyl at a level of 2 ppm, though acetaldehyde is also thought to contribute to the flavour (Lindsay *et al.*, 1965). The role of the products of citrate metabolism in ripened cheeses is less clear; diacetyl may be important in Cheddar cheese by its synergic action with other compounds produced in the maturation stage (Manning & Robinson, 1973). However, good Cheddar cheese can be made without citrate-fermenting starter bacteria (Lawrence & Pearce, 1972).

3. LIPIDS AS SOURCES OF FLAVOUR

Milk triglycerides are quantitatively the most important lipid class in cheese. Phospholipids are also present as components of the fat globule membrane and bacterial structures, but they are not thought to contribute flavour compounds to maturing cheese (Law *et al.*, 1973). Triglycerides are degraded more extensively in some cheese types than others and their contribution to cheese flavour profiles varies correspondingly.

For example, the Italian cheeses, Romano, Parmesan and Provolone, are entirely dependent for their distinctive flavour on free fatty acids released by deliberately added animal lipases (Moskowitz, 1980). Free fatty acids are also vital for the development of typical flavour in blue-vein cheeses, both as flavour compounds in their own right and as substrates for oxidation to methyl ketones, the second group of key flavour compounds. The milk or curds of the blue-vein varieties are inoculated with spores of blue-green *Penicillium* moulds (*P. roqueforti*). When the freshly made cheeses are 'spiked' to admit air, the spores germinate and the resulting mould growth spreads throughout the inside of the cheese, aided by an open texture conferred by the inclusion of gas-forming leuconostocs in the starter culture (Moreau, 1980).

Mould growth is partly controlled by the NaCl concentration in the cheese (Kinsella & Hwang, 1976). Salt is normally applied dry to the outer surface from which it diffuses to the cheese interior. Godinho & Fox (1981a,b) showed that *P. roqueforti* grew most quickly in the middle depth region, rather than the rind or centre regions because high and low salt concentrations were inhibitory, whereas intermediate concentrations (1–3% w/w) were stimulatory. The spores of *P. roqueforti* germinate early in maturation and mycelia are visible after 8–10 days. Maximum growth is reached by 1–3 months. The mycelia and spores are strongly lipolytic, with two lipases being secreted by the mould, but only the acid enzyme is significant in cheese ripening (Monassa & Lamberet, 1982). High NaCl concentrations inhibit lipolysis (Godinho & Fox, 1981c) but in zones of cheese whose NaCl concentration favours good mycelium growth, lipolysis proceeds relatively quickly.

The rate of release of fatty acids by the lipases of *Penicillium* sp. governs the rate of ketone formation (Fan *et al.*, 1976). Methyl ketones are formed from fatty acids, by partial oxidation via the fatty acid β-oxidation pathway which occurs in both spores and mycelium. The β-oxo-acyl-CoA, formed by enzymic dehydrogenation of fatty-acyl β-hydroxyacyl-CoA derivatives is deacylated by thiohydrolase to form free β-keto acid. The decarboxylase which then catalyses formation of corresponding methyl ketones is most active in mycelia, and β-oxolauric acid (yielding 2-undecanone) is its preferred substrate, but the preponderance of heptan-2-one in blue cheese is due to the preference of the preceding thiohydrolase for β-oxo-octanyl-CoA. The white surface mould of Brie and Camembert cheese (*P. camemberti*) also generates free fatty acids and oxidizes them to methyl ketones which are essential for the correct flavour balance of these varieties. Oxidative activity is variable and the main products are 2-nonanone and 2-undecanone (Dumont & Adda, 1979; Lamberet *et al.*, 1982).

The role of lipolysis and free fatty acids (FFA) is more difficult to assess in cheeses whose ripening does not involve mould growth. Surface smear floras of yeasts, micrococci and/or *B. linens* on semi-soft cheeses tend to be lipolytic and probably contribute to flavour development (Devoyod, 1969; Sorhaug & Ordal, 1974). Many hard and semi-hard ripened cheeses rely on their lactic flora for flavour development and these organisms are probably capable of limited lipolysis (Stadhouders & Veringer, 1973; Umemoto & Sato, 1975; Paulsen *et al.*, 1980). These varieties probably owe part of their characteristic flavour to low concentrations of volatile free fatty acids (5–10 µg g^{-1} cheese). For example, amounts of these acids, other than acetic acid, increase during Cheddar cheese maturation due to weak esterase and lipase activities of the milk flora and the starter streptococci (Stadhouders & Veringer, 1973). Although these volatile fatty acids are included in many synthetic flavour formulations (Law, 1983), evidence for their involvement in typical Cheddar cheese flavour/aroma is equivocal and contradictory. Studies with enzyme-modified cheese suggest that increased lipolysis by food grade enzymes in American Cheddar increases its flavour intensity favourably provided that high, rancidity-inducing lipase levels are avoided (Sood & Kosikowski, 1979). Experience at NIRD, however, suggests that increasing levels of FFA above those normally found in cheese at any particular age only impairs flavour balance and does not increase the intensity of typical flavour (Law, 1982; Law & Wigmore, 1984). The normal levels of fatty acids found in cheese (approximately 500 ppm) represent amounts well above typical flavour and aroma thresholds, ranging from 0·3–100 ppm (Baldwin *et al.*, 1973) so that they would be expected to contribute to the overall organoleptic qualities of Cheddar cheese. However, claims that fatty acids are important because low-fat or fat-free cheeses do not develop flavour (Ohren & Tuckey, 1969) are oversimplistic since deviations from an optimum fat percentage in the product alters its characteristics markedly and may have indirect effects on flavour retention and perception. Fat-free 'Cheddar' is so unlike normal Cheddar as to be irrelevant to the discussion.

Patton (1963) claimed that volatile fatty acids (C_2–C_8) were the 'backbone' of Cheddar aroma because blocking agents for carboxylic functional groups impaired the aroma of cheese fat distillates. However, Manning & Price (1977) argued that side-reactions could have occurred in these experiments which would leave the results open to different interpretations. These investigators showed that the removal of volatile fatty acids from Cheddar cheese head space did not affect its aroma at all and concluded that these acids were only important in the background taste of the cheese. Further evidence against the importance of fatty acids comes from the analysis of New Zealand Cheddars of which twenty-three out of forty-one contained no acids higher than

C_4 (Lawrence, 1967). At present we can only conclude that mixtures of alkanoic acids with carbon chains from C_2 to C_8 or C_{10} can impart cheese-like flavours either to naturally maturing cheese or in flavour mixtures for process cheese, but that their contribution to the aroma and the special character of Cheddar cheese is unproven.

Other fat-derived flavour compounds implicated in Cheddar flavour include ketones and lactones. The odd-numbered methyl ketones do not appear to be vital flavour compounds since they are absent from mature flavoured experimental cheeses made with only starter bacteria (Law, 1982). Pentanone concentrations in normal cheese are a good index of cheese age (Manning, 1979) but the compound is not necessarily involved in flavour (Aston & Dulley, 1982). Butanone is normally present in Cheddar cheese and was cited by Scarpellino as a component of desirable flavour (Law, 1982) but it is never found at concentrations higher than its threshold and tends to disappear as cheese ages.

Although lactones have been shown to improve blue cheese flavour (Jolly & Kosikowski, 1975), their contribution to Cheddar flavour is less clear and no direct evidence exists linking lactone concentrations in maturing cheese with flavour quality or intensity (Wong *et al.*, 1975). However, they are regarded as important by some flavourists, as evidenced by their inclusion in synthetic cheese flavour formulations (Law, 1983). Their formation in cheese could be as a result of spontaneous ring closure in δ-hydroxy acids released from milk glycerides by lipolysis (Boldingh & Taylor, 1962). These authors also suggested that some yeasts and moulds could reduce δ-keto acids to the hydroxy acids and convert the latter to lactones. The oxidative mechanism elucidated by Vajdi *et al.* (1979) is unlikely to occur in most cheeses because the environment is too reducing.

Esters of fatty acids are easily formed in cheese, both by microbial esterases (Morgan, 1976) or by simple chemical reactions taking place slowly over long storage periods. Since ethanol is the most abundant alcohol in cheese, ethyl esters are quite common, those of hexanoic acid often being associated with the fruity flavour defect. Esterification of short-chain fatty acids with methanethiol (normally present in cheese as a degradation product of methionine) by the surface flora of smear cheeses (*Br. linens* and micrococci; Cuer *et al.*, 1979) generates thio-esters with cheesy aromas, and such compounds could be formed non-enzymically in other cheeses. Bosch *et al.* (1982) have described the chemical preparation of a series of such thio-esters and those of propionic, butyric, valeric and octanoic acids had aroma properties compatible with their possible participation in cheese flavour development.

4. PROTEINS AS SOURCES OF FLAVOUR

4.1. Proteolysis

Like lipolysis, proteolysis in cheese proceeds to different extents and at different rates according to the composition of the cheese microflora. For example, six-month old Cheddar, with its relatively simple secondary flora, has approximately 3% of its total N content present as free amino acid N (Law *et al.*, 1976), whereas the equivalent figure for mould-ripened blue cheese can be up to 10% (Kinsella & Hwang, 1976). Such high amino acid concentrations may help in buffering cheese at relatively high pH values, thus aiding more extensive enzymic degradation of milk constituents.

The role of the amino acids and peptides released by proteolytic enzymes in hard and semi-hard cheeses is not clearly understood but they may act directly by contributing to flavour. Biede & Hammond (1979) claimed that large peptides give brothy background flavours in Swiss cheese and that proline and small peptides (in combination with Ca^{2+} and Mg^{2+}) contribute towards sweet flavours. However, the range of flavour notes which such compounds can contribute is very limited (Kirimura *et al.*, 1969). Nevertheless, the water soluble fraction of ageing cheese, of which the amino acids and peptides are an important part, is the most significant in relation to the intensity of cheese flavour according to McGugan *et al.* (1979).

Superimposed on the taste properties of these products of proteolysis may be the effect of decreasing polypeptide size on the binding of flavour compounds. McGugan *et al.* (1979) suggested that such a phenomenon may explain the greater effect of flavourless, fat-free cheese residue from old cheese on the flavour intensity of reconstituted cheese fractions. The idea is not without precedent in relation to fatty acid retention in Swiss cheese (Biede, 1978) and binding of volatile compounds to soy proteins (Fujimaki *et al.*, 1971).

The contribution of starter bacteria to this important process of cheese proteolysis varies in significance according to the extent to which secondary mould, surface, or lactic floras produce proteinases and peptidases during cheese maturation. Microbial proteolysis is also superimposed on the action of chymosin, the coagulant used in the cheese vat to destabilize casein micelles and form the gel from which the cheese curds are made. In the short term, chymosin is very bond specific (see Chapter 1) but during the following weeks and months of maturation this enzyme contributes to gross proteolysis in cheese. It is thought to produce a large proportion of the larger peptides to a lower limit of 1400 molecular weight (O'Keefe *et al.*, 1976; Desmazeaud & Gripon, 1977; Visser, 1977). Early rennet proteolysis is typified by the hydrolysis of the Phe_{23}–Phe_{24} bond (Hill *et al.*, 1974; Gripon *et al.*, 1975) or the Phe_{24}–Val_{25} bond (Creamer & Richardson, 1974) of α_{s1}-casein. Beta-casein degradation occurs only slowly in cheese and its products appear later in cheese maturation; the most sensitive bonds are Ala_{189}–Phe_{190} and Leu_{191}–Tyr_{192}.

Endopeptidases from the starter bacteria also contribute to gross casein hydrolysis (i.e. hydrolysis leading to increased pH 4·6-soluble N). For example, an intracellular endopeptidase from *Str. lactis* var. *diacetylactis* rapidly hydrolyses α_{s1}-casein (but not whey proteins) and appears to be specific for peptide bonds involving the α-amino group of hydrophobic residues (e.g. X-Leu or X-Phe; (Desmazeaud & Zevaco, 1976)). Zevaco & Desmazeaud (1980) later demonstrated that this proteinase also hydrolysed Pro_{186}–Ile_{187} and Ala_{189}–Phe_{190} in β-casein although activity was low. On the other hand, the enzyme efficiently degraded peptides derived from β-casein by chymosin action, by attacking the Lys_{176}–Ala_{177}, Lys_{119}–Val_{120} and Pro_{206}–Ile_{207} bond. These authors concluded that the starter proteinases functioned in cheese chiefly by degrading those peptides released from casein by chymosin. This is consistent with reports by several research groups that cheeses made with chymosin alone contain pH 4·6-soluble N but very little peptide and amino acid N, whereas cheeses made with starter and chymosin contain relatively large amounts of free amino N. Several independent studies (Reiter *et al.*, 1969; O'Keefe *et al.*, 1976; Visser, 1977) have established that in both Gouda and Cheddar cheeses the proteinases of starter bacteria can slowly degrade whole casein to low molecular weight (<1400) peptides and amino acids. In normal cheese, chymosin-mediated gross proteolysis is so rapid (especially of α_{s1}-casein) that it is doubtful whether this activity is significant. The most important role of starter proteinases and peptidases appears to be the degradation of large rennet-derived peptides to small peptides and amino acids. Recent evidence from studies with *prt*⁻ variants suggests that the cell-bound extracellular proteinases of starters are significant in cheese proteolysis (Mills & Thomas, 1980) but a proportion of the proteinase activity, and most of the peptidase activity, is intracellular and only released into the cheese matrix during the early ripening stage when the cells lyse (Law *et al.*, 1974). These intracellular peptidases have been widely studied and described in numerous papers (Law & Kolstad, 1983). For example, starch gel zymograms reveal the presence of dipeptidases in *Str. lactis* capable of hydrolysing a wide range of substrates containing alanine, leucine, phenylalanine, tryptophan, histidine and glycine. Similar techniques with polyacrylamide gel electrophoresis (PAGE) revealed the presence of further amino- and dipeptidases in *Str. lactis, diacetylactis* and *cremoris* active against peptides and peptide derivatives containing arginine, methionine, tyrosine and proline. An amino-peptidase from *Str. cremoris* has hydrolytic activity which is greatest on peptides released by chymosin from α_{s1}-casein. Some of these peptidases have been purified and characterized. Collectively, the peptidases of Group N streptococci are probably capable of completely hydrolysing casein to free amino acids after initial proteolysis by chymosin and starter proteinases. Despite their predominantly neutral or alkaline pH optima, these enzymes retain sufficient activity in cheese to function during maturation and can be recovered in active form from extracts of ageing cheese (Cliffe & Law, 1979).

The contribution of thermophilic starter bacteria to cheese proteolysis has not been studied to the same extent as that of mesophilic starters, but they are known to produce a range of proteinases and peptidases. Lactobacilli tend to be more proteolytic than *Str. thermophilus*, though the activity of both groups is rather weak (Tourneur, 1972). Little is known about proteinase specificity; the surface bound proteinase is most active against β-casein (Argyle *et al.*, 1976; Chandan *et al.*, 1982). Proteinases and peptidases are produced by *Lb. casei* and the specificities of the peptidases are known (Desmazeaud & Juge, 1976; Eggiman & Bachmann, 1980). They hydrolyse a variety of di- and tripeptides containing lysine, leucine, tryptophan, phenylalanine, histidine, methionine and proline. Both the carboxypeptidase and endopeptidase have very restricted specificities for synthetic substrates and it is impossible at present to assess their contribution to cheese proteolysis. The endopeptidase has subsequently been identified as a ribosomal arylpeptidyl amidase (El Soda & Desmazeaud, 1981) and such enzymes have now been detected in many other species of *Lactobacillus* (El Soda *et al.*, 1982). The proteolytic systems of lactic-acid bacteria were reviewed recently (Law & Kolstad, 1983).

Compared to the action of lactic-acid bacteria, the white surface mould *P. camemberti* and the internal blue mould *P. roqueforti* contribute a much greater part of the proteolytic activities to the cheeses to which they are added. The action of the proteinases and peptidases of *P. camemberti* in texture formation and amino acid liberation in Brie and Camembert have received much research attention (Gripon & Debest, 1976; Lenoir *et al.*, 1979; Trieu-Cuot *et al.*, 1982a,b).

Two extracellular proteinases (acid aspartyl and neutral metalloproteinase) have been purified and characterized; the peptidases include two distinct carboxypeptidases and an aminopeptidase. The *P. camemberti* proteinases cleave bonds in α_{s1}- and β-casein, producing large peptides. Even though few points in the polypeptide chains of these proteins are cleaved (4 in α_{s1}-casein and 3 in β-casein) the effect on cheese texture is profound because the network of associating casein molecules is broken down very effectively (Creamer & Olson, 1982). Trieu-Cuot & Gripon (1982) have proposed a sequence for proteolysis in Camembert cheese which involves chymosin, plasmin (native milk proteinase) and the two *P. camemberti* proteinases. Early α_{s1}-casein proteolysis is thought to be chymosin-mediated but the similar actions of the *Penicillium* acid enzyme and chymosin prevent an assessment of their relative importance beyond the 7th day of maturation, when mould growth and enzyme secretion become significant near the cheese surface. Because chymosin-mediated hydrolysis of β-casein is slow in cheese, its degradation by the mould enzymes is more easily observed from the 7th day of maturation. Lenoir & Auberger (1982) suggested that the acid proteinase activity was highest in young cheese from 6 to 16 days after manufacture, but it disappeared after 22 days. The neutral proteinase, on the other hand, became active from the 13th day and persisted at a high level for at least one month. However, from their detailed observations of casein-derived peptides, Trieu-Cuot & Gripon (1982) have come to the opposite conclusion, suggesting that the aspartyl acid proteinase is more important than the neutral metalloproteinase for most of the maturation period. Since they have directly observed the appearance of specific products of both *Penicillium* proteinases, the evidence of Trieu-Cuot & Gripon is the more convincing at present. Plasmin also hydrolyses β-casein but its action is not seen until 21–35 days, after the pH of the cheese has risen to nearer the neutral/alkaline optimum of this enzyme.

The relationship between the proteinases and the peptidases in the sequential breakdown of caseins to amino acids in the cheese is not clear because relatively little is known about the specificities of the latter group of enzymes. Both extra- and intracellular enzymes are produced; presumably the intracellular peptidases only become significant in older cheeses in which the *Penicillium* mycelia become moribund and leak their enzymes into the cheese matrix. In any case the pH optima of these peptidases are relatively high (Auberger *et al.*, 1982) so that they could only be effective in the later stages of maturation. Despite the lack of agreement on the role of yeasts in the microflora development and ripening of Camembert cheese, Schmidt (1982) has shown that many isolates produce a range of intracellular proteolytic enzymes capable of liberating small peptides and amino acids from casein. They may therefore contribute to the final stages of protein breakdown together with the *P. camemberti* peptidases.

4.2. Amino Acid Catabolism

Amino acid catabolism by the surface flora yields a variety of flavour compounds and precursors, thought to be important to the development of the subtle, distinct flavour and aroma of Camembert cheese. For example, ammonia contributes to the aroma profile and results from amino acid deamination by the contaminating yeast microflora, particularly species of *Geotrichium* (Greenburg & Ledford, 1979; Hemme *et al.*, 1982). Cheeses which also develop a growth of *Brevibacterium linens* are likely to take on a stronger ammoniacal aroma since this organism deaminates the amino groups of most amino acids including the side chain groups (Hemme *et al.*, 1982). General flavour characteristics common to many surface ripened varieties of cheese have been attributed to 3-methyl-1-butanol, phenylethanol and phenol; though direct evidence is lacking for mechanisms operating in cheese, it is known that many microorganisms, including *Br. linens* can convert leucine, phenylalanine and tyrosine respectively into these three compounds. Volatile sulphur compounds make up another important group in soft cheese flavour and aroma. *P. camemberti* itself was reported to produce H_2S, dimethylsulphide and methanethiol from methionine (Tsugo & Matsuoka, 1962) by a combination of oxidative deamination and demethiolation and *Br. linens* also has this capability (Sharpe, *et al.*, 1977). Methanethiol has no proven direct role in Camembert aroma but it may be involved in non-enzymic reactions such as that with formaldehyde to yield bis-(methylthio)-methane, thought to be important to the aroma profile (Dumont *et al.*, 1976; Cuer *et al.*, 1979). *Br. linens* also appears to interact with micrococci to produce a variety of thioesters, but these are thought to be of greater significance in surface smear cheese (Cuer *et al.*, 1979).

Volatile sulphur compounds derived from the end products of proteolysis may also contribute to flavour in Cheddar cheese and related varieties, though the amounts formed in these cheeses are lower than those in mould or smear-ripened cheeses. For example, methanethiol is measured in $ng\,g^{-1}$ in Cheddar cheese, but in $\mu g\,g^{-1}$ in mould-ripened cheese. It is present in the Cheddar-like aroma fraction of cheese (Manning & Robinson, 1973) and its concentration in the headspace of ageing cheese or cheese containing added reducing agents correlates very closely with flavour intensity scores, irrespective of the age of the cheese (Manning *et al.*, 1976; Manning, 1979; Green & Manning, 1982). The absence of methanethiol from cheese, or its removal from headspace volatiles, coincides with the absence of typical flavour and aroma respectively (Manning, 1974; Manning & Price, 1977). More recent evidence appears to throw some doubt on the pivotal flavour role of methanethiol as proposed by Manning *et al.* (1976) in that Aston & Douglas (1983) interpreted the correlation between flavour intensity and methanethiol as being coincidental. Also Lamparsky & Klimes (1981) were unable to detect methanethiol at all in young Cheddar cheese whose flavour was judged to be mild but typical. On balance, the weight of evidence remains in favour of a role for methanethiol in maturing cheese flavour, but its mechanism of formation in Cheddar cheese is not certain. Very few Cheddar cheeses contain methanethiol-producing coryneform bacteria (Law & Sharpe, 1978) and although many raw milk pseudomonads produce this volatile from methionine, the enzymic reaction does not appear to proceed significantly in cheese. Although methanethiol is found in cheeses made with starter, but not in chemically acidified cheese, the starters themselves do not produce it directly (Law & Sharpe, 1978). Since it has been observed that the redox potential of starter cheese is much lower (-150 to $-200\,mV$) than that of chemically acidified cheese ($+300\,mV$; Law *et al.*, 1976), it has been postulated (Law & Sharpe, 1977; Manning, 1978) that methanethiol is generated by non-enzymic reactions, but only remains stable in low redox cheese made with starter. This idea has been confirmed by studies with reducing agents (Green & Manning, 1982) but the role of the thiol in flavour could not be confirmed because the artificially reduced cheeses possessed strong flavour defects.

Manning (1979) and Green & Manning (1982) studied possible

chemical mechanisms for methanethiol formation in cheese and concluded that it is closely linked with the release of hydrogen sulphide. The latter compound is usually present in cheese and probably contributes to flavour, though the concentration at which it is present is probably only critical if it is too high, when it can be detected as a flavour defect (Manning, 1979). Hydrogen sulphide can be produced by lactobacilli under the carbon-limited low pH conditions in cheese (Sharpe & Franklin, 1962) and Manning (1979) showed that it could react, by an addition or substitution mechanism, with casein or methionine to yield methanethiol. This, and other possible non-enzymic mechanisms, were discussed at length by Green & Manning (1982).

4.3. Proteolysis and Flavour Defects

As well as contributing to desirable flavour, products of proteolysis can cause off-flavours. In particular, peptides have been identified as the source of the bitter defect in cheese. Opinions on the size of bitter peptides vary but it is generally agreed (Ney, 1971; Matoba & Hata, 1972; Guigoz & Solm, 1976; Champion & Stanley, 1982) that they contain a high proportion of hydrophobic amino acids (e.g. leucine, phenylalanine, proline). Of the casein fractions, α_{s1}-casein is regarded as the main source of bitter peptides (Richardson & Creamer, 1973; Adda *et al.*, 1982). Many factors govern whether or not sufficient quantities of bitter peptides accumulate in ripening cheese to exceed flavour thresholds and manifest the defect. They include the extent of rennet retention in the curd, the numbers of starter bacteria present in the young cheese, the peptide-degrading capacity of the starter cells, and the rate at which they lyse and release the peptidases.

Opinions differ as to the importance of proteolysis by mesophilic starters in the production of bitter defects in cheese. Early hypotheses suggested that bitter peptides were produced by chymosin and that the so-called 'bitter' starters were those which had insufficient peptidase activity to break down the bitter peptides to non-bitter peptides and amino acids (Czulak, 1959). However, the situation is more complex than this. While it is true that chymosin produces bitter peptides from casein, the starter proteinases can also do this and, indeed, can produce small bitter peptides from non-bitter, casein-derived peptides. Lowrie & Lawrence (1972) and Lowrie *et al.* (1974) suggested that this latter process is the single most important determinant in bitterness development and that starters which multiplied at relatively high cooking temperatures during Cheddar manufacture (the 'fast' starters) were the most likely to give bitter cheese, simply because the resultant high cell numbers contributed large quantities of bitter peptide-producing proteinases. This hypothesis was supported with experimental evidence showing that 'bitter' starters could be made to produce non-bitter cheese if their number in curds was restricted by controlled bacteriophage infections or by higher cooking temperatures. Conversely, the slow 'non-bitter' starters made bitter cheese if they were allowed to multiply to high cell numbers by altering the manufacturing process. Direct evidence for the involvement of starter cell wall proteinases in the development of bitterness was provided recently by the observation that proteinase-deficient variants of 'fast' starters produce less bitterness in cheese than their parent strains, even when total starter cell populations are high (Mills & Thomas, 1980).

The factors controlling bitter defects in Gouda cheese appear to be more complex since the starters generally reach high populations in curd at the relatively low cooking temperatures used for this variety. Stadhouders & Hup (1975) showed that factors influencing the retention of chymosin in Gouda curd (e.g. cooking temperature, initial milk pH) also influence the tendency of the cheese to become bitter. They emphasized that some starter strains produce more bitter peptide-degrading peptidases than others. It is not known whether these are specific peptidases confined to non-bitter strains or general peptidases present at different levels. Chiba & Sato (1980) identified both dipeptidase and amino-peptidase activity in fractions of cell-free extracts from starter streptococci capable of reducing bitterness, but individual enzymes were not isolated. It appears, then, that proteolysis by mesophilic starters is important in producing the bitter defect in cheese but its contribution depends on the cheese variety in question.

5. CONCLUSIONS

Ripening mechanisms and flavour compounds are reasonably well defined in strong-flavoured cheeses, including blue-vein, surface mould and surface smear varieties, with very active, complex microfloras. This is generally because a few classes of flavourful compounds are present in sufficiently high concentrations to be identified and organoleptically evaluated against a relatively low background of minor constituents. The important compounds are more difficult to pick out in the case of other semi-hard and hard cheeses whose predominantly lactic floras are less well endowed with degradative enzymes. So many compounds can be identified at low, yet significant concentrations that it is impossible to select those which are important. It is probably the case that the flavour profiles of these varieties are composed of many different interacting compounds whose individual contribution to desirable flavour is not discernible experimentally. The emphasis on cheese flavour research has traditionally fallen on the volatile compounds but there is a growing realization that the non-volatile, water soluble products of proteolysis are very important in determining the strength of cheese taste.

REFERENCES

ACCOLAS, J. P., VEAUX, L., VASSAL, L. & MOQUOT, G. (1978) *Le Lait* **58**, 118.

ACCOLAS, J. P., HEMME, D., DESMAZEAUD, M. J., VASSAL, L., BOUILL-ANNE, C. & VEAUX, M. (1980) *Le Lait* **60**, 487.

ADDA, J., GRIPON, J.-C. & VASSAL, L. (1982) *Food Chemistry* **9**, 115.

ARGYLE, P. J., MATHISON, G. E. & CHANDAN, R. C. (1976) *J. Appl. Bacteriol.* **41**, 175–84.

ASTON, J. W. & DOUGLAS, K. (1983) *Australian J. Dairy Technol.* **38**, 66.

ASTON, J. W. & DULLEY, J. R. (1982) *Australian J. Dairy Technol.* **37**, 59.

AUBERGER, B., MONTALS, M. & LENOIR, J. (1982) In *Proc. XXI Int. Dairy Congress*, Moscow, Vol. 1, Book 1. Mir Publishers, Moscow, p. 276.

BABEL, F. J. (1977) *J. Dairy Sci.* **60**, 815.

BALDWIN, R. E., CLONINGER, M. R. & LINDSAY, R. C. (1973) *J. Food Sci.* **38**, 528.

BIEDE, S. L. (1978) *Dissertation Abstr.* **38**, 3110-B.

BIEDE, S. L. & HAMMOND, E. G. (1979) *J. Dairy Sci.* **62**, 227.

BOLDINGH, J. & TAYLOR, R. J. (1962) *Nature, London* **194**, 909.

BOSCH, S., VAN DEN LAND, E. V. & STOFFELSMA, J. (1982) *US Patent* 4,332,829.

BOYOVAL, P. & DESMAZEAUD, M. J. (1983) *Le Lait* **63**, 187.

CHAMPION, H. M. & STANLEY, D. W. (1982) *Canadian Inst. Food Sci. Technol. J.* **15**, 283.

CHANDAN, R. C., ARGYLE, P. J. & MATHISON, G. E. (1982) *J. Dairy Sci.* **65**, 1408.

CHIBA, Y. & SATO, Y. (1980) *Japanese J. Dairy Food Sci.* **29**, 161.

CLIFFE, A. J. & LAW, B. A. (1979) *J. Appl. Bacteriol.* **47**, 65.

CREAMER, L. K. & OLSON, N. F. (1982) *J. Food Sci.* **47**, 631.

CREAMER, L. K. & RICHARDSON, B. C. (1974) *New Zealand J. Dairy Sci. Technol.* **9**, 9.

206   *Barry A. Law*

CUER, A., DAUPHIN, G., KERGOMARD, A., DUMONT, J.-P. & ADDA, J. (1979) *Agric. Biol. Chem.* **43**, 1783.

CZULAK, J. (1959) *Australian J. Dairy Technol.* **14**, 177.

CZULAK, J., HAMMOND, L. A. & HORWOOD, J. F. (1974) *Australian J. Dairy Technol.* **29**, 124.

DESMAZEAUD, M. J. & GRIPON, J.-C. (1977) *Milchwissenschaft* **32**, 731.

DESMAZEAUD, M. J. & JUGE, M. (1976) *Le Lait* **56**, 241.

DESMAZEAUD, M. J. & ZEVACO, C. (1976) *Annales de Biologie animale Biochimie et Biophysique* **16**, 851.

DEVOYOD, J. J. (1969) *Le Lait* **49**, 20.

DUMONT, J.-P. & ADDA, J. (1979) In *Progress in Flavour Research,* D. G. Land & H. E. Nursten, Eds. Applied Science Publishers, London, p. 255.

DUMONT, J.-P., PRADEL, G., ROGER, S. & ADDA, J. (1976) *Le Lait* **56**, 18.

EGGIMAN, B. & BACHMANN, M. (1980) *Appl. Environ. Microbiol.* **40**, 876.

EL SODA, M. & DESMAZEAUD, M. J. (1981) *Agric. Biol. Chem.* **45**, 1693.

EL SODA, M., ZEYADA, N., DESMAZEAUD, M. J., MASHALY, R. & ISMAIL, A. (1982) *Sciences Alimentaires* **2**, 261.

FAN, T. Y., HWANG, D. H. & KINSELLA, J. E. (1976) *J. Agric. Food Chem.* **24**, 443.

FRYER, T. F. (1982) *Proc. XXI Int. Dairy Congress,* Moscow, Vol. 1, Book 1. Mir Publishers, Moscow, p. 485.

FUJIMAKI, M. S., ARAI, S. & YAMASHITA, M. (1971) *Proc. Int. Symp. on Conversion & Manufacture of Foodstuffs by Microorganisms,* Kyoto. Saikon Publishing Company, Tokyo, p. 19.

GODINHO, M. & FOX, P. F. (1981a) *Milchwissenschaft* **36**, 205.

GODINHO, M. & FOX, P. F. (1981b) *Milchwissenschaft* **36**, 329.

GODINHO, M. & FOX, P. F. (1981c) *Milchwissenschaft* **36**, 457.

GREEN, M. L. & MANNING, D. J. (1982) *J. Dairy Res.* **49**, 737.

GREENBURG, R. S. & LEDFORD, R. A. (1979) *J. Dairy Sci.* **62**, 368.

GRIPON, J.-C. & DEBEST, B. (1976) *Le Lait* **56**, 423.

GRIPON, J.-C., DESMAZEAUD, M. J., LE BARS, D. & BERGERE, J.-L. (1975) *Le Lait* **55**, 502.

GUIGOZ, Y. & SOLM, S. J. (1976) *Chem. Senses & Flavour* **2**, 71.

HEMME, D., BOUILLANNE, C., MÉTRO, I. & DESMAZEAUD, M. J. (1982) *Sciences Alimentaires* **2**, 113.

HETTINGA, D. H. & REINBOLD, G. W. (1972) *J. Milk Food Technol.* **35**, 358.

HILL, R. D., LEHAR, E. & GIVOL, D. (1974) *J. Dairy Res.* **41**, 147.

JOLLY, R. C. & KOSIKOWSKI, F. V. (1975) *J. Agric. Food Chem.* **23**, 1175.

KINSELLA, J. E. & HWANG, D. (1976) *Biotechnol. Bioengineer.* **18**, 927.

KIRIMURA, J., SHIMIZU, A., KIMIZUKA, T., NINOMIYA, T. & KATSUYA, N. (1969) *J. Agric. Food Chem.* **17**, 689.

KLETER, G. (1976) *Netherlands Milk Dairy J.* **30**, 254.

LAMBERET, G., AUBERGER, B., CANTERI, C. & LENOIR, J. (1982) *Revue Laitière Française* **406**, 13.

LAMPARSKY, D. & KLIMES, I. (1981) In *Proc. Weurman Symp.* **3**, P. Schreier, Ed. Walter de Gruyter & Co., Berlin, p. 557.

LAW, B. A. (1982) In *Fermented Foods, Economic Microbiology,* A. H. Rose, Ed., Vol. 7. Academic Press, London, p. 147.

LAW, B. A. (1983) *Perfumer & Flavorist* **7**, 9.

LAW, B. A. & KOLSTAD, J. (1983) *Antonie van Leeuwenhoek* **49**, 225.

LAW, B. A. & SHARPE, M. E. (1977) *Dairy Industries Int.* **42**, 10.

LAW, B. A. & SHARPE, M. E. (1978) *J. Dairy Res.* **45**, 267.

LAW, B. A. & WIGMORE, A. S. (1984) *J. Dairy Res.* **51** (In press).

LAW, B. A., SHARPE, M. E., CHAPMAN, H. R. & REITER, B. (1973) *J. Dairy Sci.* **56**, 716.

LAW, B. A., SHARPE, M. E. & REITER, B. (1974) *J. Dairy Res.* **41**, 137.

LAW, B. A., CASTANON, M. J. & SHARPE, M. E. (1976) *J. Dairy Res.* **43**, 301.

LAWRENCE, R. C. (1967) *New Zealand J. Dairy Sci. Technol.* **2**, 55.

LAWRENCE, R. C. & PEARCE, L. E. (1972) *Dairy Industries* **37**, 73.

LENOIR, J. & AUBERGER, B. (1982) In *Proc. XXI Int. Dairy Congress,* Moscow, Vol. 1, Book 1. Mir Publishers, Moscow, p. 336.

LENOIR, J., AUBERGER, B. & GRIPON, J.-C. (1979) *Le Lait* **59**, 244.

LINDSAY, R. C., DAY, E. A. & SANDINE, W. E. (1965) *J. Dairy Sci.* **48**, 863.

LOWRIE, R. J. & LAWRENCE, R. C. (1972) *New Zealand J. Dairy Sci. Technol.* **7**, 51.

LOWRIE, R. J., LAWRENCE, R. C. & PEBERDY, H. F. (1974) *New Zealand J. Dairy Sci. Technol.* **9**, 116.

McGUGAN, W. A., EMMONS, D. B. & LARMOND, E. (1979) *J. Dairy Sci.* **62**, 398.

MANNING, D. J. (1974) *J. Dairy Res.* **41**, 81.

MANNING, D. J. (1978) *Dairy Industries Int.* **43**, 37.

MANNING, D. J. (1979) *J. Dairy Res.* **46**, 523.

MANNING, D. J. & PRICE, J. C. (1977) *J. Dairy Res.* **44**, 357.

MANNING, D. J. & ROBINSON, H. M. (1973) *J. Dairy Res.* **40**, 63.

MANNING, D. J., CHAPMAN, H. R. & HOSKING, Z. D. (1976) *J. Dairy Res.* **43**, 313.

MATOBA, T. & HATA, T. (1972) *Agric. Biol. Chem.* **36**, 1423.

MILLS, O. E. & THOMAS, T. D. (1980) *New Zealand J. Dairy Sci. Technol.* **15**, 131.

MONASSA, A. & LAMBERET, G. (1982) In *Proc. XXI Int. Dairy Congress,* Moscow, Vol. 1, Book 1. Mir Publishers, Moscow, p. 509.

MOQUOT, G. (1979) *J. Dairy Res.* **46**, 113.

MOREAU, C. (1980) *Le Lait* **60**, 254.

MORGAN, M. E. (1976) *Biotechnol. Bioengineer.* **18**, 953.

MOSKOWITZ, G. J. (1980) In *The Analysis and Control of Less Desirable Flavours in Foods and Beverages*, G. Charalambous, Ed. Academic Press, New York, p. 53.

NEY, K. H. (1971) *Z. Lebensmittel-Unterschung und Forschung* **147**, 337.

OHREN, J. A. & TUCKEY, S. L. (1969) *J. Dairy Sci.* **52**, 598.

O'KEEFE, R. B., FOX, P. F. & DALY, C. (1976) *J. Dairy Res.* **43**, 97.

PATTON, S. (1963) *J. Dairy Sci.* **46**, 856.

PAULSEN, P. V., KOWALEWSKA, J., HAMMOND, E. G. & GLATZ, B. A. (1980) *J. Dairy Sci.* **63**, 912.

PURKO, M., NELSON, W. D. & WOOD, W. A. (1951) *J. Dairy Sci.* **34**, 699.

REITER, B., SOROKIN, Y., PICKERING, A. & HALL, A. J. (1969) *J. Dairy Res.* **36**, 65.

RICHARDSON, B. C. & CREAMER, L. K. (1973) *New Zealand J. Dairy Sci. Technol.* **8,** 46.

SCHMIDT, J. L. (1982) In *Proc. XXI Int. Dairy Congress,* Moscow, Vol. 1, Book 1. Mir Publishers, Moscow, p. 365.

SHARPE, M. E. & FRANKLIN, J. G. (1962) *Proc. 8th Int. Congress Microbiol.* **B.11.3,** 46.

SHARPE, M. E., LAW, B. A., PHILLIPS, B. A. & PITCHER, D. G. (1977) *J. General Microbiol.* **101,** 345.

SOOD, V. K. & KOSIKOWSKI, F. V. (1979) *J. Dairy Sci.* **62,** 1865.

SORHAUG, T. & ORDAL, Z. J. (1974) *Appl. Microbiol.* **25,** 607.

STADHOUDERS, J. & HUP, G. (1975) *Netherlands Milk Dairy J.* **29,** 335.

STADHOUDERS, J. & VERINGER, H. A. (1973) *Netherlands Milk Dairy J.* **27,** 77.

STEFFEN, C. (1979) *Milk Industry* **81,** 24.

THOMAS, T. D., ELLWOOD, D. C. & LONGYEAR, V. M. C. (1979) *J. Bacteriol.* **138,** 109.

TODESCO, R., LODI, R., MUCCHETTI, G. & CARINI, S. (1981) *Il Latte* **6,** 741.

TOURNEUR, C. (1972) *Le Lait* **52,** 149.

TRIEU-CUOT, P. & GRIPON, J.-C. (1982) *J. Dairy Res.* **49,** 501.

TRIEU-CUOT, P., ARCHIERI-HAZE, M.-J. & GRIPON, J.-C. (1982a) *J. Dairy Res.* **49,** 487.

TRIEU-CUOT, P., ARCHIERI-HAZE, M.-J. & GRIPON, J.-C. (1982b) *Le Lait* **62,** 234.

TSUGO, T. & MATSUOKO, H. (1962) In *Proc. XVI Int. Dairy Congress,* Copenhagen, Vol. B, p. 385.

TURNER, K. W. & THOMAS, T. D. (1980) *New Zealand J. Dairy Sci. Technol.* **15,** 265.

UMEMOTO, Y. & SATO, Y. (1975) *Agric. Biol. Chem.* **39,** 2115.

VAJDI, M., NAWAR, W. W. & MERRITT, C. (1979) *J. American Oil Chem. Soc.* **56,** 906.

VISSER, F. M. W. (1977) *Netherlands Milk Dairy J.* **31,** 210.

WONG, N. P., ELLIS, R. & LA CROIX, D. E. (1975) *J. Dairy Sci.* **58,** 1437.

ZEVACO, C. & DESMAZEAUD, M. J. (1980) *J. Dairy Sci.* **63,** 15–24.

Chapter 8

The Accelerated Ripening of Cheese

BARRY A. LAW

National Institute for Research in Dairying, Shinfield, Reading, UK

1. INTRODUCTION

Traditionally, cheese manufacturers have accepted that for any given cheese variety there must be a certain time lag between the initial conversion of milk, and the realization of the value of the resulting cheese. However, while some maturation time is inevitable, present understanding of some aspects of cheese ripening has led to much experimentation into means of shortening it by speeding up the reactions which generate flavour and modify texture. Almost all of the reported attempts to accelerate cheese ripening fall into one of four categories shown in Table 1; all have accompanying advantages and disadvantages.

Elevated temperatures offer the simplest approach to speeding up cheese ripening from a technical standpoint, and there is no legal constraint on changing normal storage temperatures. However, while this approach may give satisfactory results with the highest quality cheese, with a pH, moisture, salt concentration and bacteriological quality close to ideal, any tendency towards lower standards would probably be exaggerated by a high storage temperature. The economic loss resulting from down-grading or complete rejection of a proportion of a factory's output could easily outweigh any savings in storage costs for those cheeses which were suitable for flavour acceleration.

Enzyme addition offers a more specific alternative method for speeding up flavour-producing reactions. For example, proteolysis and lipolysis are important processes in the maturation of most cheese varieties and there are a number of cheap commercial food grade

TABLE 1
Major Categories of Methods for Accelerating Cheese Ripening

Method	Advantages	Disadvantages
Elevated temperature	No legal barriers, technical simplicity	Non-specific action, increased microbial spoilage potential
Enzyme addition	Low cost, specific action, choice of flavour options	Limited sources of useful enzymes, danger of over-ripening, difficult to incorporate, legal barriers
Modified starter	Probably no legal barriers, natural enzyme balance retained, easy to incorporate	Technical complexity, uneconomical at present
Slurry	Very rapid flavour development	High microbial spoilage potential, final product requires processing

enzymes which can be used to achieve their selective acceleration. There are at present a number of difficulties associated with the use of enzymes which have yet to be overcome. For example, their addition to cheese is not legally acceptable at present, the range of useful enzymes commerically available is limited, and some enzymes are difficult to control in terms both of their addition to cheese and their action in cheese.

The starter bacteria used for cheesemaking contain many important enzymes, particularly the peptidases responsible for much of the amino acid-releasing activity in cheese (Chapter 7). The use of attenuated starters in addition to normal cultures in cheesemaking can result in higher concentrations of such enzymes in cheese, without concomitant overproduction of lactic acid in the vat. Other possible modifications include genetic manipulations to introduce new enzyme producing capabilities, or overall changes in culture composition to achieve high starter populations late in the cheesemaking process. Advantages of such methods include the ease of incorporation of the starter cells into cheese curd, the retention of a natural enzyme balance and the decreased likelihood of legal objections. On the debit side, the preparation of modified starters is technically and scientifically complex and at present starter cultures are expensive to grow.

Liquid slurry methods have the main advantage of speed of flavour development but they are difficult to stabilize against microbiological spoilage and the end product cannot be used directly.

The following discussion will consider these four methods of accelerated cheese ripening in more detail.

2. ELEVATED TEMPERATURE

Most cheese varieties are ripened at relatively low temperatures (<15°C) though some may be subject to periods in 'warm' rooms to develop specific characteristics (e.g. Emmental for 'eye' development). When cheese is regarded as a means of preserving milk nutrients for later consumption, low storage temperatures are logical and beneficial, minimizing the risk of spoilage. However, modern creameries apply stringent standards of hygiene to both the raw material and the process. The use of elevated ripening temperatures is therefore worthy of consideration under these conditions of reduced spoilage risk.

Most reports on the use of temperature control to accelerate cheese ripening have involved hard and semi-hard cheeses with relatively simple microfloras. Attempts to increase flavour in low fat Gouda cheese showed that flavour balance was easily impaired; at 16°C proteolysis was accelerated more than lipolysis and the cheeses developed the bitter defect (International Dairy Federation, 1983). High temperature forced ripening of Edam cheese has been reported to cause microbiologically induced texture and flavour spoilage due to *Clostridium tyrobutyricum*, propionibacteria and heterofermentative lactobacilli (International Dairy Federation, 1984).

Accelerated ripening of Cheddar cheese by the application of elevated temperatures has met with some success. For example, Law *et al.* (1979) noted that among several manufacturing parameters including starter strain, vat design or ripening temperature, it was the last factor which had the greatest influence on the flavour intensity of the cheeses after 6 and 9 months storage. In one case cheeses stored for 6 months at 13°C had a mean flavour intensity score of 4·4 (medium/mature on the 0–8 scale) while those stored for 9 months at 6°C scored only 3·2 (mild on the same scale). There was no significant difference in defect scores overall but the cheeses stored at 13°C had a lower incidence of bitterness than those stored at 6°C, suggesting a differential temperature optimum for peptide-degrading enzymes compared with proteinases in favour of the former. The data in Table 2 are taken from

Barry A. Law

TABLE 2
Accelerated Cheddar Cheese Ripening by Elevated Storage Temperature
(Data sources: Law *et al.* (1979), Law & Wigmore (1982b) and Aston *et al.*
(1983b))

Normal temperature (°C)	Experimental temperature (°C)	Ripening period (weeks)	Advancement[a] (weeks)
6	12	8	8
6	13	24	24
6	18	8	12
8	20 and 13	4(20°C) + 4(13°C)	4·4
8	20 and 13	4(20°C) + 20(13°C)	3

[a] Based on extrapolation of data for low-temperature cheese to estimate time taken to reach a similar flavour intensity score as high-temperature cheese.

Law *et al.* (1979), Law & Wigmore (1982b) and Aston *et al.* (1983b) to illustrate in more detail the degree of flavour advancement which can be achieved by the temperature control method. The work of Aston *et al.* (1983b) suggests that this method of fast ripening only gives a significant advantage over a relatively short period, though Law *et al.* (1979) reported continuing advantage up to 24 weeks. The reasons for such differences are not clear.

The studies reported above all deal with cheese made on a pilot scale under strictly controlled experimental conditions. In anticipation that non-starter lactic-acid bacteria (NSLAB) in commercial cheeses may grow too quickly at high maturation temperatures and cause flavour and texture defects, Fryer (1982) investigated a new temperature regime in extensive factory-scale trials.

The method depends on cheeses being pressed quickly and transferred to blast-cooled stores so that the cheese temperature falls rapidly to <10°C. If the cheeses are then held for 14 days, there is a continuous slow growth of NSLAB in such a way that they ferment residual lactose only to lactic acid and not to any of the alternative products which could produce off-flavours. The conditions should be such that the count of NSLAB is $<10^3 \text{ g}^{-1}$ at hooping and remains $<10^6 \text{ g}^{-1}$ after 14 days. Fryer (1982) claims that after this initial period the balanced flavour of the cheese can be encouraged to develop more rapidly at a relatively high storage temperature, without fear of secondary bacterial metabolism causing flavour defects. The method has the advantage of simplicity, provided the manufacturer has the necessary equipment for rapid pressing, and should be applicable to any hard or semi-hard cheese whose flavour development depends on enzymic processes, as distinct from the metabolism of a viable secondary flora. However, the assumption is made that only NSLAB can grow and influence the quality of the cheese; this requires verification at elevated temperatures and it would be prudent to determine whether normally dormant microorganisms begin to multiply in warmer cheeses.

3. ENZYME ADDITIONS

Since many cheese varieties are matured by the action of indigenous enzymes, rather than by their viable microflora, it seems logical to think of accelerating their maturation by artificially increasing the concentration of certain of these enzymes where they have a definite role in the ripening process.

3.1. Proteinases
Proteolysis is such a centrally important (if poorly understood) part of flavour and texture development in cheese, that many attempts have been made to influence events with exogenous microbial proteinases, ranging in origin from the starter streptococci to fungi. There is little doubt that accelerating the production of low molecular weight peptides and amino acids by adding commercial food grade proteinases produces strong-flavoured cheese in a short time. For example, Sood & Kosikowski (1979c) used acid and neutral proteinases to halve the normal ripening time of American Cheddar cheese. However, careful choice of enzymes is necessary if flavour imbalance and bitter defects are to be avoided. Sood & Kosikowski (1979b) suggested a screening method to determine suitable rates of addition, using rapidly-ripened semi-liquid cheese slurries. Unfortunately much of the published literature is lacking in definitive organoleptic and rheological data on the effects of exogenous proteinases during cheese storage. Law & Wigmore (1982a,b) used a trained flavour panel to generate numerical data suitable for statistical analysis. Also, changes in cheese texture were evaluated instrumentally. The results of the flavour evaluations are shown in Table 3.

TABLE 3

Effect of Commercial Proteinases on Flavour Development in Cheddar Cheeses After Two Months' Maturation (Data from Law & Wigmore, 1982b)

Proteinase	Concentration (units g^{-1})[a]	[b] Cheddar intensity (0–8 scale)	[b] Bitter intensity (0–4 scale)	[b] Other off-flavours (0–4 scale)
Acid (A. oryzae)	50	1·8*	3·3	1·1
	10	2·6	2·8	0·7
	20	2·7	2·6	0·4
	0·4	2·1	2·2	0·3
	0·08	2·2	0·7	0·2
Neutral (B. subtilis)	50	3·1*	1·6	0·9
	10	3·4*	0·7	0·4
	2	3·2*	0·1	0·2
	0·4	2·7	0·1	0·2
Alkaline (B. licheniformis)	2	2·9	3·5	0·0
Pronase (Str. griseus)	2	4·1*	0·5	0·6
Untreated	—	2·4	0·1	0·2

[a] One unit = amount of enzyme required to bring about an increase in A_{595} of 0·5 in 15 min using the Hide Powder Azure (HPA) assay of Cliffe & Law (1982), 1 HPA unit = 10^{-5} Anson units; [b] mean score from 24 panelists; * significantly different from untreated cheese at $p < 0.05$.

The aspartyl acid proteinase of *Aspergillus oryzae* always produced bitter cheese without significant enhancement of typical flavour, even when addition levels were varied over a range of two orders of magnitude. Bacterial neutral proteinase (*Bacillus subtilis*) also produced bitter cheese at high levels of addition, but an optimum amount can be determined which significantly enhances the intensity of flavour in Cheddar cheese without producing this flavour defect. Addition of the enzyme below this level did not improve the flavour significantly. There seems to be no reason why this type of enzyme should not be effective in any hard or semi-hard cheese whose maturation is partially proteolytic in nature. The *B. licheniformis* alkaline proteinase (subtilisin) produced very bitter cheese when added at the same level as neutral proteinase and was not investigated further because of its excessive cost. Pronase is a broad-specificity proteinase from *Streptomyces griseus* with weak aminopeptidase activity. It produced strong flavours in cheese but the cheese was bitter. At lower levels of addition no flavour enhancement was produced (Law & Wigmore, unpublished data).

All proteinase treatments resulted in soft-bodied, crumbly cheese, though this was least noticeable where neutral proteinase had been added. For example, a subjective panel assessment (based on the procedure of Green *et al.*, 1981) suggested that only one of three replicates treated with neutral proteinase was significantly more crumbly than untreated control cheese (Law & Wigmore, 1982a). The textural differences appeared to be due to excessive breakdown of β-casein by the exogenous proteinases.

Treatment of cheese with *B. subtilis* neutral proteinase has increased potential if used in combination with temperature control. For example, a flavour intensity equivalent to normal 4 month cheese can be produced in one month by holding the cheese at 18°C (Law & Wigmore, 1982b). Because this enzyme shows little activity below 8°C, low temperatures can be used to slow down the rate of maturation if market fluctuations reduce the demand for mature cheese. Further protection against over-ripening by *B. subtilis* proteinase is provided by its inherent instability under the conditions prevailing in cheese (Law & Wigmore, 1982a). Cheese treated with the enzyme, then stored at 6°C for 2 months did not mature significantly more quickly than untreated cheese if it was subsequently transferred to a ripening room at 12°C, indicating that the proteinase was no longer active. Acid proteinases, on the other hand, remain active for long periods in cheese, partly explaining the excessive proteolysis which they catalyse during ripening.

Attempts to increase desirable cheese flavours by proteinase treatment are usually limited by the onset of defects. Law & Wigmore (1982a) found that it was possible to increase amounts of small peptides and amino acids by up to 400% of the normal level in 2 month-old cheese, but any enzyme treatment which resulted in increases greater than approximately 150% also gave bitter cheese with an excessively meaty flavour. This problem can be overcome by combining the predominantly endopeptidase activity of neutral proteinase with an exopeptidase preparation. As well as yielding a high proportion of low molecular weight N with low degrees of gross proteolysis (high amino acid : peptide ratio), this treatment also allows for a reduction in dose rates

TABLE 4

Effect of Combined Commercial Proteinase and Peptidase-rich *Streptococcus lactis* Extract on Flavour Development and Proteolysis in Cheddar Cheese (Data from Law & Wigmore (1983))

Treatment	[a] *Cheddar cheese flavour intensity (0–8 scale)*	*Proteolysis (TCA-soluble N as % of control)*[b]	*Amino acid:peptide ratio (SSA-soluble N/ TCA-soluble N)*[c]
Str. lactis extract	2·9*	100	4·0
Proteinase	3·1*	140	2·9
Str. lactis extract + proteinase	3·8**	200	4·4
No treatment (control)	2·2	100	2·3

[a] Mean score from 24 panelists; [b] TCA = trichloracetic acid; [c] SSA = sulphosalicylic acid; * significantly different from control at $p < 0.05$; ** significantly different from control at $p < 0.01$.

of proteinase so that body/texture defects can be minimized. The results of a trial carried out by Law & Wigmore (1983) with *B. subtilis* proteinase and a peptidase-rich extract of *Str. lactis* are shown in Table 4. It is interesting to note that the introduction of additional peptidase activity alone into the cheese has a relatively minor impact on flavour intensity and on proteolysis. This suggests that gross proteolysis is rate-limiting in the sequence of reactions leading to the accumulation of small peptides and amino acids.

A similar approach to accelerated ripening was described by Kalinowski *et al.* (1979) except that they used *Penicillium roqueforti* or *P. candidum* proteinase with an extract of *Lactobacillus casei* or *Str. lactis*, which was also attributed with proteinase activity. The lactic-acid bacteria probably contributed peptidase rather than proteinase to the system. Treatment of cheese milk with such enzyme combinations was said to halve the ripening time of Edam or Tilsit cheese, without changing their typical flavour characteristics.

3.2. Lipases

The acceleration of lipolysis by the addition of either animal or microbial lipases has been successfully applied to the relatively strong-flavoured cheeses. For example, although Italian hard cheeses are traditionally made with lipases, their flavour can be further enhanced or modified using commercial lipases in addition to the more usual rennet paste-associated lipases or lamb esterase (Bottazzi, 1965; Moskowitz, 1980). Blue cheese ripening can also be enhanced with a lipase preparation from *Aspergillus* sp. (Jolly & Kosikowski, 1975). This treatment has the combined effect of increasing the typical fatty acid flavour, and catalysing more rapid liberation of lactone and ketone precursors (cf. Chapter 7, Section 3).

Lipases have been beneficial in promoting the ripening of Egyptian cheese varieties. El Salim *et al.* (1978) succeeded in halving the ripening time of Ras cheese by adding commercial gastric lipases to the cheese milk. Baky *et al.* (1982a,b) reported that rapidly ripened slurries of Ras cheese could be prepared by including a *Mucor michei* esterase in the formulation in order to accelerate the liberation of fatty acids. The slurries were then added to Ras cheese to enhance its flavour. Egyptian soft pickled cheese, like its European equivalents, benefits from fatty acid flavour notes and it is not surprising that lipases are reported to shorten its maturation time (El Salim *et al.*, 1981).

The efficacy of lipases as agents for rapid ripening of Cheddar and related types is open to interpretation, perhaps because the notion of 'typical mature flavour' has changed as this type of cheese has been made and consumed in more and more countries. American Cheddar is thought to benefit from the cautious addition of lipases to the curd (Sood & Kosikowski, 1979a) and volatile fatty acids are certainly part of the flavour profile of rapidly ripened cheese pastes for use in processed cheese. However, attempts to accelerate the development of typical flavour in English Cheddar cheese using commercial lipases have failed (Law & Wigmore, 1984); the lipases were screened for their ability to release either short- or long-chain fatty acids in order to differentiate between their effects on flavour. The long-chain (C_{12}–C_{16}) fatty acids released by a *Mucor meihei* lipase produced an unpleasant 'soapy' flavour defect, while the short-chain acids released by animal esterases produced an unclean flavour. Many levels of addition were investigated but these enzymes either produced no flavour effect at all or they produced defects; no compromise could be reached whereby desirable flavour could be enhanced without defects. Even when they were added together with proteinases, the lipase did not accelerate the formation of typical flavour.

3.3. β-Galactosidase

This enzyme catalyses the formation of galactose and glucose from lactose and therefore has no direct role in producing flavour compounds. However, it has been claimed (Anon, 1977) that both cheese starter bacteria and the secondary floras grow better in milk and cheese which has been pre-treated with β-galactosidase (lactase) so that flavour development is enhanced indirectly. This original claim has come under critical scrutiny since then, partly due to the lack of definitive evidence as to the relative efficiency with which cheese microorganisms utilize lactose on the one hand, and glucose and galactose on the other. Gilliland et al. (1972) considered glucose to be a preferred energy source compared with lactose for the mesophilic starters and this would explain the apparent stimulation of acid production during Cheddar cheesemaking with lactose-hydrolysed milk (Thompson & Brower, 1976; Marschke & Dulley, 1978). Both groups of workers reported some acceleration of flavour development in their experimental cheeses, corresponding to reductions of about 10–50% in maturation times, dependent on the amount of lactase added to the original milk. The faster ripening was accompanied by more rapid proteolysis, measured as low molecular weight soluble nitrogen fractions. However, more recent evidence suggests that the apparent stimulation of acid production by the starter, and the accelerated flavour production, were due to contaminating proteinase activities in the commercial lactase preparations (Marschke et al., 1980; Grieve et al., 1983a). Marschke et al. (1980) demonstrated that the lactase in Maxilact (Kluyveromyces lactis β-galactosidase) could be heat inactivated leaving the proteolytic activity in the preparation intact. When it was found that the heated material was almost as effective as the original in accelerating cheese ripening and protein breakdown, they concluded that the proteinase was the most important factor involved. Grieve et al. (1983a) have partially characterized three proteolytic enzymes in Maxilact and K. lactis autolysates. They found an acid endopeptidase, serine endopeptidase and a carboxypeptidase.

Hemme et al. (1979) also concluded that proteolytic action, rather than lactose hydrolysis, was the mechanism by which Maxilact stimulated growth and lactic acid production by thermophilic starter cultures. Neither Lb. helveticus nor Lb. bulgaricus was stimulated in Maxilact-treated milk, but Str. thermophilus grew better than it did in untreated milk. This stimulation was shown to be due to products of proteolysis; none of the culture components was stimulated by pure β-galactosidase or by added glucose or galactose.

Claims by Gooda *et al.* (1983) that starter and secondary flora stimulation, and not contaminating proteinase, are responsible for improved Cheddar flavour development in cheeses made with Maxilact-treated milk may keep the debate alive but are not well supported by experimental evidence. No data were given on the numbers of bacteria in curds and ripening cheeses, and the extensive data on proteolysis does not help in differentiating degradation due to starter enzymes or to contaminating proteinase activity. Also, the claim that high numbers of viable starter bacteria in curd can accelerate cheese flavour development is not supportable on present evidence; both Lowrie & Lawrence (1972) and Law *et al.* (1979) have previously noted that cheeses in which starters had multiplied to relatively low numbers ($\sim 10^8 \, \mathrm{g}^{-1}$) during manufacture subsequently developed better and stronger flavour than those in which starter numbers in curds exceeded $10^9 \, \mathrm{g}^{-1}$, partly due to the excessive bitterness associated with high levels of starter-mediated proteolysis in the latter cheese.

3.4. Enzyme Incorporation into Cheese

The formation of cheese from milk involves complex physico-chemical changes which are not completely understood (see Chapter 1). However, it is reasonable to suggest that exogenous enzymes would be incorporated homogenously into cheese only if they were mixed with the cheesemilk and therefore present as the milk gel was formed. Full dispersal of enzymes added at a later stage would probably be limited by their high molecular weight, and correspondingly low rate of diffusion. On the other hand, although enzymes added to the milk are likely to be better distributed, only a proportion is retained in the curd, and a significant amount is always lost into the whey at the stage of separation. This represents an economic loss of the enzyme itself, and it may also render the whey unsuitable for other uses. For example, whey contaminated with lipases cannot be considered for ice cream manufacture because the product contains fats which would be susceptible to degradation. Similar problems could arise in the case of proteolytic contamination where the whey was being used as a source of functional proteins in baked confectionary products. Proteinases added to milk also present a more direct threat to the economics of cheesemaking in that they can attack the milk proteins in the vat. As a result, some of the protein is lost as low molecular weight breakdown products into the whey. This obviously leads to reduced yields of cheese from a relatively expensive raw material. On an experimental scale, enzyme-treated cheese can be made by adding the dried enzyme to the curd before it is pressed and packed. This is usually done by 'diluting' the enzyme powder in the dry salt which is a normal ingredient of most cheeses. This step is necessary because amounts of enzyme are usually very small in relation to the curds to be treated (Sood & Kosikowski, 1979c; Law & Wigmore, 1982a). On this scale, the enzyme can be handled safely and distributed reasonably evenly, though Law & Wigmore (1982a) noted some evidence of uneven texture resulting from this method of addition. On a factory scale, powdered enzymes are less likely to be accepted because of potential health risks to personnel (particularly from inhalation) and distribution difficulties arising from the relatively large curd particle size compared to that encountered during experimental cheesemaking. Spraying or injection of liquid enzyme concentrates has been considered but can give severe storage problems, especially when proteinases are used. Other technical difficulties encountered with these techniques have been reviewed by Law (1980).

Enzyme encapsulation offers the possibility of overcoming all of the problems outlined above. For example, proteinases can be kept separate from milk proteins by the capsule material during the vat stage of cheesemaking. Also, suitably sized and charged capsules could be designed for maximum retention within the curd matrix as it formed during milk coagulation. Once entrapped, the encapsulated enzyme would be as well distributed as any other milk constituents and, when released from the capsules, in intimate contact with its substrates.

This approach has been investigated by a number of workers. Schafer (1975) encapsulated lipases in formaldehyde-treated gelatin but encountered difficulties in melting the capsule material to release the enzyme into Mozzarella cheese at the designed temperature (45°C). Magee & Olson (1981a,b) used milk fat to entrap cell-free extracts of *Str. diacetylactis* together with substrates and cofactors for diacetyl production. The capsules apparently maintained the integrity of the enzyme sequence for the pathway from pyruvate to diacetyl when incorporated into directly acidified low fat cheese; the corresponding unencapsulated additions produced eight times less diacetyl in cheese than the encapsulated system. The use of fat capsules has been extended to the formation of acetic acid by cell-free extracts of *Gluconobacter oxydans* (Braun *et al.*, 1982). The system initially included partially purified alcohol dehydrogenase, ethanol and NAD, but more recently Braun & Olson (1983) have found that extracts of *Str. lactis* var. *maltigenes* can also be encapsulated to recycle reduced

NAD (NADH) by using it in the conversion of leucine to 3-methylbutanal and 3-methylbutanol. Such flavour-generating systems have potential applications in the intensification of flavour in low fat or directly acidified cheese products. Further developments reported by Rippe *et al.* (1983) indicate that mature Cheddar cheese flavour may be reached more quickly with encapsulated enzymes from *Pseudomonas putida* which produce methanethiol from methionine in a protected environment. The chief limitation of these capsules is their instability at cheese cooking temperatures. Since they melt above 33°C they are unsuitable for Cheddar and Emmental and related types and their use in Edam and Gouda cheese would be difficult to accommodate. As Magee & Olson (1981a) pointed out in their description of the methodology, higher melting fractions of milk fat may solve this problem.

4. MODIFIED STARTERS

These fall into two main categories. In the first, the starter bacteria themselves remain unmodified, but the preparation conditions are changed so that they produce more metabolites which contribute towards the desirable properties of the cheese. In the second category, the starter bacteria are modified either physically, chemically or genetically so that their enzyme balance is changed.

4.1. Culture modifications
Sovietski cheese (an Emmental-type cheese made in the Soviet Union) can be produced by accelerated ripening techniques involving specially grown starter preparations combined with elevated temperatures (Dilanian, 1980). This, as well as other Soviet hard and semi-hard varieties, can be ripened more quickly with starter supplements containing very high numbers of cells. For example, the growth of aroma-producing bacteria in cheese can be stimulated by using 'hydrolysed bacterial starter' (HBS).

A typical preparation would be made initially from a 3·5% inoculum of mesophilic starter in high heat-treated milk. In order to break down the milk proteins to low molecular weight starter nutrients (amino acids and peptides) the culture is treated with xg rennet powder according to the formula:

$$x\text{g} = \frac{2 \cdot 5}{\substack{\text{starter} \\ \text{dose (litres)}}} \text{ per 100 litres cheesemilk for final make}$$

The HBS containing rennet is incubated at 24–28°C for 20–24 h, then homogenized and used at 0·6–1·0% (v/v) to inoculate the cheesemilk. The resulting cheese is said to ripen more quickly (by an unspecified amount) than those made with normal starter culture (Dilanian, 1980). Yeasts which do not form alcohol have also been used to stimulate the flora of Soviet hard cheeses, including Emmental types. They do so by virtue of their capacity to release free amino acids and to produce vitamin growth factors. As an example, starter cultures made up of 0·2% lactobacilli, 0·4% streptococci and 5×10^3 yeast cells ml^{-1} gave reductions of up to 25% in ripening times in cheesemaking trials (Dilanian, 1980).

Claims for accelerated making and maturation of Emmental, Gouda and Carre de l'Est cheese are made for a neutral proteinase preparation from *Micrococcus caseolyticus* ('Rulactine'; Vassal *et al.*, 1982). The proposed use of the enzyme is similar in principle to that described by Dilanian (1980) for HBS, in that Rulactine is used to treat a small volume of milk prior to cheesemaking in order to liberate peptides and amino acids as growth stimulants for the starter bacteria. The following is a typical user scheme for this enzyme: to treat a 10 000 litre vat, the Rulactine/milk digest is prepared the previous day by mixing 1 litre sterile milk with 90 g of enzyme (10 000 units g^{-1}) and incubating at 20°C overnight. The digest is then added to the cheese vat as it is being filled. As a result of the action of Rulactine on casein the starter grows and produces acid more quickly in the vat, thus speeding up the make process. The growth of secondary bacteria is also said to be stimulated. Residual Rulactine is also assumed to accelerate protein breakdown in the maturing cheese. The chief benefit for the manufacturer is said to come from a more rapid development of typical cheese texture (even for soft cheese); claims on flavour enhancement are vague and not supported by any data at present.

4.2. Cell Function Modifications

Law (1980) reviewed the available literature covering physical and chemical modifications; no significant advances have been reported since then. To summarize, methods in this general category have involved lysozyme treatment and heat shock, both of which prevent the bacteria from producing acid, while having a minimal effect on their proteolytic and peptide-degrading enzymes. Heat shock methods (e.g. Pettersson & Sjostrøm, 1975) are potentially useful and favoured over the lysozyme method for technical simplicity. The technique has been successfully applied to accelerated ripening of Swedish cheese on an experimental scale (Petterson & Sjostrom, 1975). Exterkate (1979) showed that acid production by starters could also be prevented by treating them with *n*-butanol. This had the added effect of activating some of the peptidases in the cell by up to 10 times their normal levels. However, the effect of these cells on cheese ripening has not yet been reported.

Genetic variants of lactic streptococci which are unable to produce lactic acid from lactose can be isolated from most starter cultures. They can also be produced by treating cells with agents which cause the loss of plasmid DNA (see Chapter 4). A number of these *lac⁻* variants are now well characterized (Gasson, 1983; Grieve *et al.*, 1983b) and they have obvious potential in accelerated ripening techniques which require high concentrations of degradative enzymes from the starter bacteria as potential flavour producers.

A detailed study of *lac⁻* variants as agents of accelerated cheese ripening was described recently in a series of papers (Grieve & Dulley, 1983; Aston *et al.*, 1983a,c) in which a *lac⁻prt⁻* (proteinase-deficient) strain of *Str. cremoris* C2 was added to cheeses made with several different normal single strain cultures of *Str. cremoris*. Numbers of the variant retained in the cheese curds were between 2 and 10 times the numbers of the viable normal starter and although flavour advancement was reported, it was not proportional to the variant populations. Cheeses containing the variant matured more quickly than normal in terms of the rate of proteolysis during storage, and a good correlation was reported between free amino acid concentrations and overall flavour intensity. However, the effects of the variants on typical Cheddar cheese flavour are difficult to interpret. Although commercial graders judged the experimental cheeses to have been advanced in perceived age by between 1 and 3 months in 6 months (Grieve & Dulley, 1983) the panel scores for typical, rather than overall flavour, showed that little advantage was gained by using the variants (Aston *et al.*, 1983c). The authors reported that the variants imparted an atypical flavour to the cheese which may have accounted for the higher overall intensity scores compared with control cheese made with normal starter alone. This is consistent with the findings of Law & Wigmore (unpublished data) that *lac⁻* variants of *Str. lactis* NCDO 712 only alter Cheddar cheese by imparting a flavour described by the panel as 'oxidized' and 'fruity'. More encouraging results have been obtained with *lac⁻ Str. cremoris* in that the flavour of treated cheeses develops more quickly with fewer atypical flavour notes.

Other genetic modifications to starter bacteria can apparently enhance their proteolytic and lipolytic activity. For example, Dilanian *et al.* (1976) described the use of starters containing X-ray mutants of lactobacilli which had increased proteolytic activities. The mutants were selected for cheesemaking by their ability to liberate more free amino acids (particularly glutamic acid, alanine, tyrosine, methionine and leucine) from milk than did their corresponding parent strains. Armianski cheeses containing the mutants ripened more quickly than control cheeses, as judged by the greater accumulation (up to 25%) of non-protein N-fractions. Mature flavour developed in 45 instead of the normal 60 days (Dilanian & Sarkisyan, 1970) but the claimed increase in cheese quality is impossible to interpret in the absence of more detailed data. A method for the production and isolation of X-ray mutants of *Lb. casei* was described by Singh & Ranganathan (1978) but their effect in cheese was not reported. Singh *et al.* (1981a,b) later reported that UV mutants of *Str. diacetylactis* and X-ray mutants of *Lb. bulgaricus* and *casei* produced more carbonyl compounds and long-chain free fatty acids when incubated in milk for 48 h. The usefulness of such enhanced properties requires further clarification, however, since not all carbonyls are beneficial to cheese flavour and long-chain fatty acids are usually associated with 'soapy' flavour defects.

Genetically modified starters will probably take on a greater significance when suitable methods of genetic manipulation have been developed so that specific, desirable enzymes or metabolic pathways can be introduced into them. Also, our increasing knowledge of the genetics of starter bacteria (Chapter 4) should, in the foreseeable future, provide the possibility of making strains which selectively overproduce enzymes such as peptidases.

5. SLURRY METHODS

Kristoffersen *et al.* (1967) developed a method of accelerating flavour development in Cheddar cheese curd by increasing its moisture content and incubating at 30°C. The slurries, containing about 40% total solids, developed strong flavours in days rather than months. However, despite extensive studies on the effects of aeration, curd milling acidities and additives (e.g. reduced glutathione, riboflavin) by Singh

& Kristoffersen (1972) and Harper & Kristoffersen (1970) the mechanism by which flavour is so quickly produced remains unclear and the process is difficult to control. In particular the growth of yeasts causes off-flavour development, but if potassium sorbate is added clean flavours can be produced (Dulley & Taylor, 1972). Dulley (1976) used slurried curd to accelerate cheese ripening; the slurries were prepared from normally manufactured Cheddar curd by aseptically blending it with 5% NaCl and 3% sorbate until a smooth semi-liquid paste was obtained. This was then stored at 30°C in closed containers for 7 days, then incorporated into cheese by addition either to the cheesemilk, the curd before Cheddaring or the salted curd before pressing. Addition at the latter stage avoided loss of slurry into the whey. Cheese moisture was increased 2% by the addition of 6% (w/w) slurry (Dulley, 1976) and this in itself resulted in accelerated flavour development, but the slurry appeared to have a greater, additional effect. Although the overall flavour scores of the slurry-containing cheeses were depressed by a higher incidence of off-flavours compared with controls, they developed flavour intensities in 4, 12 and 24 weeks equivalent to normal cheeses aged 1·8, 3·5 and 5·1 weeks older, respectively. In contrast with other acceleration methods, the use of slurry did not appear to increase proteolysis (TCA-soluble N) in cheese. The author considered that the high numbers of lactobacilli (10^5–10^7 cfu g^{-1}) were responsible for accelerating ripening.

It would be interesting and relevant to investigate the ability of the lactobacilli in the slurry-containing cheeses to produce specific compounds previously cited as contributing to Cheddar cheese flavour (e.g. H_2S, methanethiol; see Chapter 7).

Von Bockleman & Lodin (1974) also reported that increasing the numbers of lactobacilli from 10^6 to 10^9 g^{-1} in cheese (by adding slurried mature cheese to milk for Swedish Prastost manufacture) produced a stronger flavour than normal cheese after 3 months ripening. However, the effects of the treatment on the experimental cheeses were not reported in detail. Baky *et al.* (1982a) similarly adapted the slurry method to the ripening of Egyptian Ras cheese by adding it to the cheese curd. The ripening time was reduced by half and proteolysis and lipolysis were accelerated, but the mechanism involved was not made clear. Presumably the slurry contained high numbers of proteolytic and lipolytic bacteria which had developed from the natural flora of the slurried curd while it was incubated at 30°C, before being added to curd for normal cheesemaking.

Sutherland (1975) investigated the use of ripened Cheddar cheese slurries as replacements for the mature component of processed cheese. He was able to make an acceptable product from normal slurries, 'sulphury' (glutathione-treated) slurries, and lipase-treated slurries. Sood & Kosikowski (1979a) carried out similar trials with enzyme-treated ultrafiltration retentates in processed cheese and claimed some improvement over the conventional product.

6. CONCLUSIONS

The lack of detailed knowledge of cheese flavour compounds, and the mechanisms which generate them during maturation, remains the main obstacle to the rational, scientific design of accelerated cheese ripening systems. However, despite such limitations a number of promising developments have emerged based on the control of maturation temperature, and the use of exogenous enzymes. In the latter field, the goal is to provide 'packaged' enzyme systems for the complete control of cheese ripening, even to the extent of changing flavour characteristics in pre-formed cheese and decreasing the levels of flavour in older cheese, as the market demands. Reactions which are not normally possible in cheese because of pH, salt or redox potential, may be made to proceed in encapsulated enzyme/substrate complexes operating under optimum conditions. Recent advances in understanding the genetics of cheese starter bacteria may be expected to yield specialized strains designed not only to make cheese, but to ripen it quickly.

REFERENCES

ANON (1977) *Dairy and Ice Cream Field* **160**, 66K.
ASTON, J. W., DURWARD, I. G. & DULLEY, J. R. (1983a) *Australian J. Dairy Technol.* **38**, 55.
ASTON, J. W., FEDRICK, I. A., DURWARD, I. G. & DULLEY, J. R. (1983b) *New Zealand J. Dairy Sci. Technol.* **18**, 143.
ASTON, J. W., GRIEVE, P. A., DURWARD, I. G. & DULLEY, J. R. (1983c) *Australian J. Dairy Technol.* **38**, 59.
BAKY, A. A. A., EL-NESHEWY, A., RABIE, A. H. M. & FAHARAT, S. M. (1982a) *J. Dairy Res.* **49**, 337.
BAKY, A. A. A., EL-FAK, A. M. & RABIE, A. M. (1982b) *Dairy Industries Int.* **47**, 21.
BOTTAZZI, V. (1965) *Scienza e Technica Lattiero-Casearia* **16**, 229.

BRAUN, S. D. & OLSON, N. F. (1983) *J. Dairy Sci.* **66** (Supplement 1), 77.
BRAUN, S. D., OLSON, N. F. & LINDSAY, R. C. (1982) *J. Food Sci.* **47**, 1803.
CLIFFE, A. J. & LAW, B. A. (1982) *J. Dairy Res.* **49**, 209.
DILANIAN, Z. KH. (1980) *Principles of Cheesemaking.* Osnovy syrodeluja, Pishchevaya Promyshlennost, Moscow.
DILANIAN, Z. KH. & SARKISYAN, R. A. (1970) *Promeschi Armenii* **10**, 104.
DILANIAN, Z. KH., MAKARIAN, K. & CHUPRINA, D. (1976) *Milchwissenschaft* **31**, 219.
DULLEY, J. R. (1976) *Australian J. Dairy Technol.* **31**, 143.
DULLEY, J. R. & TAYLOR, G. C. (1972) *Proc. Australian Biochem. Soc.* **5**, 52.
EL-SALIM, M. H. ABD., EL-SHIBINY, S., EL-BAGOURY, E., AYAD, E. and FAHMY, N. (1978) *J. Dairy Res.* **45**, 491.
EL-SALIM, A. M. H., EL-SHIBINY, S., MONEIB, A. F., EL-HEIBA, A. & EL-KHAMY, A. F. (1981) *Egyptian J. Dairy Sci.* **9**, 143.
EXTERKATE, F. A. (1979) *J. Dairy Res.* **46**, 473.
FRYER, T. F. (1982) In *Proc. XXI Int. Dairy Congress* Moscow, Book 1, Vol. 1. Mir Publishers, Moscow, p. 485.
GASSON, M. J. (1983) *Antonie van Leeuwenhoek* **49**, 275.
GILLILAND, S. E., SPECK, M. L. & WOODWARD, J. R. (1972) *Appl. Microbiol.* **23**, 21.
GOODA, A., BEDNARSKI, W. & POZNANSKI, S. (1983) *Milchwissenschaft* **38**, 65.
GREEN, M. L., TURVEY, A. & HOBBS, D. G. (1981) *J. Dairy Res.* **48**, 343.
GRIEVE, P. A. & DULLEY, J. R. (1983) *Australian J. Dairy Technol.* **38**, 49.
GRIEVE, P. A., KITCHEN, B. J. & DULLEY, J. R. (1983a) *J. Dairy Res.* **50**, 469.
GRIEVE, P. A., LOCKIE, B. A. & DULLEY, J. R. (1983b) *Australian J. Dairy Technol.* **38**, 10.
HARPER, W. J. & KRISTOFFERSEN, T. (1970) *J. Agric. Food Chem.* **18**, 563.
HEMME, D., VASSAL, L., FOYEN, H. & AUCLAIR, J. (1979) *Le Lait* **59**, 597.
INTERNATIONAL DAIRY FEDERATION (1983) Document 157, 33.
INTERNATIONAL DAIRY FEDERATION (1984) (In press).
JOLLY, R. C. & KOSIKOWSKI, F. V. (1975) *J. Agric. Food Chem.* **23**, 1175.
KALINOWSKI, L., FRACKIEWICZ, E., JANISZEWSKA, L., PAWLIK, A. & KIKOLSKA, D. (1979) US Patent 4,158,607.
KRISTOFFERSEN, T., MIKOLAJCIK, E. M. & GOULD, I. A. (1967) *J. Dairy Sci.* **50**, 292.
LAW, B. A. (1980) *Dairy Industries Int.* **45**, 15.
LAW, B. A. & WIGMORE, A. S. (1982a) *J. Dairy Res.* **49**, 137.
LAW, B. A. & WIGMORE, A. S. (1982b) *J. Soc. Dairy Technol.* **35**, 75.
LAW, B. A. & WIGMORE, A. S. (1983) *J. Dairy Res.* **50**, 519.
LAW, B. A. & WIGMORE, A. S. (1984) *J. Dairy Sci.* **51** (In press).
LAW, B. A., HOSKING, Z. D. & CHAPMAN, H. R. (1979) *J. Soc. Dairy Technol.* **32**, 87.
LOWRIE, R. J. & LAWRENCE, R. C. (1972) *New Zealand J. Dairy Sci. Technol.* **7**, 51.
MAGEE, E. L. & OLSON, N. F. (1981a) *J. Dairy Sci.* **64**, 600.
MAGEE, E. L. & OLSON, N. F. (1981b) *J. Dairy Sci.* **64**, 616.

MARSCHKE, R. J. & DULLEY, J. R. (1978) *Australian J. Dairy Technol.* **33,** 139.

MARSCHKE, R. J., NICKERSON, D. E. J., JARRETT, W. D. & DULLEY, J. R. (1980) *Australian J. Dairy Technol.* **35,** 84.

MOSKOWITZ, G. J. (1980) In *The Analysis and Control of Less Desirable Flavours in Foods and Beverages,* G. Charalambous, Ed. Academic Press, New York, p. 53.

PETTERSSON, H. E. & SJOSTROM, G. (1975) *J. Dairy Res.* **42,** 313.

RIPPE, J. K., LINDSAY, R. C. & OLSON, N. F. (1983) *J. Dairy Sci.* **66** (Supplement 1), 77.

SCHAFER, H. W. (1975) *Dissertation Abstr. Int.* **36,** 1127 B.

SINGH, S. & KRISTOFFERSEN, T. (1972) *J. Dairy Sci.* **55,** 744.ͻ

SINGH, J. & RANGANATHAN, B. (1978) *Experientia* **34,** 183.

SINGH, J., CHANDER, H. & RANGANATHAN, B. (1981a) *Milchwissenschaft* **36,** 266.

SINGH, J., RANGANATHAN, B. & CHANDER, H. (1981b) *Milchwissenschaft* **36,** 742.

SOOD, V. K. & KOSIKOWSKI, F. V. (1979a) *J. Dairy Sci.* **62,** 1713.

SOOD, V. K. & KOSIKOWSKI, F. V. (1979b) *J. Food Sci.* **44,** 1690.

SOOD, V. K. & KOSIKOWSKI, F. V. (1979c) *J. Dairy Sci.* **62,** 1865.

SUTHERLAND, B. J. (1975) *Australian J. Dairy Technol.* **30,** 138.

THOMPSON, M. P. & BROWER, B. P. (1976) *Cultured Dairy Products J.* **11,** 22.

VASSAL, L., DESMAZEAUD, M. J. & GRIPON, J.-C. (1982) *Proc. Int. Symp. on Enzyme Applications in Foods,* P. Dupy, Ed. Technique et Documentation Lavoisier, Paris, p. 315.

VON BOCKLEMAN, I. & LODIN, L.-O. (1974) *Proc. XIX Int. Dairy Congress,* New Delhi **1E,** 441.

Chapter 9

Non-sensory Methods for Cheese Flavour Assessment

Donald J. Manning, Elizabeth A. Ridout and John C. Price

National Institute for Research in Dairying, Shinfield, Reading, UK

1. INTRODUCTION

Cheese is usually assessed for flavour at least twice between manufacture and marketing, initial grading being carried out between 3 and 4 weeks for the purpose of determining the most suitable market for the cheese. This assessment of the potential quality of the cheese is an essential part of the grader's duties and is of great economic importance to commercial creameries. Storage of cheese for ripening accounts for a significant part of the cost of cheese since maintaining large stores at low temperatures is expensive. The grader's aim then is to try to ensure that cheese ripened for up to 12 months will have the necessary quality for the mature cheese market.

In addition to predicting whether a cheese will ripen to a good quality mature cheese, the grader is also responsible for predicting whether a cheese will be suitable for a particular retail outlet. Even among good quality mature cheeses one can find a range of characteristic flavours and each retailer will endeavour to specify the particular type of cheese he requires for his customers.

Having selected cheeses considered suitable for long maturation, and possibly checking the cheeses during ripening, it is necessary for graders to assess the cheeses prior to despatch to the customer. This final grading is often carried out in conjunction with the buyer for the retailer and it will be at this stage that graders will know how successful was their initial grading. Cheeses which fail to come up to the required standard at this stage will either have to be used for processing or an alternative, less demanding, outlet will have to be found. Either way, the cheese will not obtain the premium price.

Before attempting to assess the reliability of present methods of assessing cheese, it is necessary to outline what the methods consist of. Grading is usually carried out under three classifications, flavour and aroma, body and texture, and colour. To assess these various aspects of quality, graders sample the cheese with a special cheese iron which enables them to remove a core from the cheese. The core still in position in the iron is then passed beneath the nose and the first assessment of the flavour of the cheese is made. Inspection of the back of the iron also provides information about how the cheese is progressing, a uniform film of fat on the back of the iron is a sure indication that from the point of view of body and texture the cheese is potentially a good one. Inspection of the core itself will provide valuable information about the body and texture and the depth and uniformity of colour. Graders rarely taste the cheese, but rubbing the cheese between finger and thumb, releasing less volatile compounds, adds to the assessment of flavour.

Although all three categories of cheese assessment are sensory, they are clearly different in their amenability to specification. Body and texture and colour are all assessed by feel and sight and their attributes can be readily expressed in descriptive terms which can be conveyed from one person to another. Flavour and aroma are quite different in this respect since descriptive terms relating to taste and aroma cannot be adequately conveyed from one person to another, unless the people concerned are trained in the same language of flavour. A great deal of work has been carried out to find ways of describing flavours, which is normally called profiling and involves describing the flavour of a product in terms of its component flavour notes. However, this is usually confined to the work of expert panels. The present specification for Cheddar at 4 weeks is as follows: body, firm but malleable; texture, close smooth and silky; colour, uniform (the colour of Cheddar can vary from cream to almost red depending upon the particular part of the UK in which the cheese is to be sold); flavour and aroma, clean and characteristic with an absence of undesirable flavours and aromas. The latter definition for flavour and aroma is particularly vague since there is no attempt to define the characteristic flavour of the cheese. Although there is a lot of evidence to show that by and large graders do an effective job, the interest shown by industry in the development of non-sensory methods for assessing cheese, clearly indicates that there is a desire for greater efficiency. To say that graders do an effective job and yet at the same time indicate a need for non-sensory methods appears to be contradictory. This is not the case; the grader effectively utilizes the limited information that can be derived from traditional methods of grading, but if this information were to be supplemented by the more detailed data available from non-sensory or instrumental methods then a greater efficiency could be achieved in the assessment of cheese quality.

Although this chapter is mainly concerned with non-sensory methods for assessing cheese flavour it was necessary to give an account of the sensory method being employed at the present in order to indicate why non-sensory methods are required and to what particular aspect of flavour they are most likely to be applied. Before leaving sensory cheese grading it is worth drawing attention to the work of McBride & Hall (1979) which showed that in tests carried out in New South Wales, Australia, involving official dairy graders, there was no overall correlation between consumer preference and dairy graders' scores for Cheddar cheese. Although there are no published results of tests of this type carried out in the UK it would seem very likely that similar results would be obtained. Without going into McBride's experiment at this stage, his findings are not entirely surprising because even if consumer surveys were carried out, it would be difficult to relate the subjective assessments of the consumers and graders. If, however, objective or instrumental methods were available, it should not be too difficult to relate the consumers' requirements to physical or chemical parameters of the cheese.

2. OBJECTIVES

The main objectives behind the development of non-sensory methods for assessing cheese are very much the same as those of the present sensory methods for grading. For mature cheese, methods are required which will give an objective assessment of the quality of the cheese and for young cheese a reliable method is required that facilitates the assessment of its potential quality. Any objective or instrumental method for assessing quality is unlikely to eliminate the need for graders but would provide information which at present is not available to graders and thereby supplement his knowledge of the cheese obtained by sensory methods. Instrumental methods would extend the grading system to enable different flavours within a variety of cheese to be characterized, thereby enabling creameries to select cheeses for a particular market or indeed to suit the different requirements of the consumer. Using instrumental methods for assessing the quality of the cheese would also mean that consumer surveys could be carried out to some advantage.

Development of methods for non-sensory assessment of cheese has almost entirely been confined to flavour and aroma. Present methods for assessing body and texture are likely to be quite adequate provided detailed specifications can be agreed upon. Nevertheless, physical methods for determining various aspects of body and texture could be used to some advantage. The approaches to non-sensory assessment of cheese quality can roughly be classified into two types: the assessment of quality by relating compositional analyses (such as moisture and salt) to cheese quality, as is being carried out by Gilles & Lawrence (1973) in New Zealand, or the determination of flavour compounds, which is the approach in the UK and Australia.

3. ASSESSMENT OF QUALITY BY COMPOSITIONAL ANALYSIS

The use of compositional analyses as an indicator of cheese quality has been pioneered by Gilles & Lawrence (1973) so successfully that in 1977 New Zealand became the first country to grade cheese in this way. By correlating compositional analyses and grading scores, Gilles & Lawrence were able to propose a system of grading based on limits for various chemical parameters. They recommended specific ranges for different parameters for first grade and premium grade Cheddar (Table 1).

Lawrence & Gilles (1980) extended their work to include the assessment of potential quality of young Cheddar cheese and concluded that this could only be achieved if rigorous controls were used on the method of manufacture, selection of starters and creamery hygiene.

Since to apply compositional analyses for grading cheese only involves standard analyses which have been carried out routinely on cheese for many years (such as salt, moisture and fat) there is little that requires saying about methodology. Nevertheless, to devise any objective method for assessing cheese quality one must inevitably relate the chemical data to sensory data, whether it be obtained from grading

TABLE 1
Recommended Chemical Composition of Premium and First Grade Cheddar

		Analyses		
Grade	% salt/ moisture	% moisture/fat free solid	% fat/water free substance	pH
Premium grade	4·0–6·0	52–56	52–55	4·95–5·1
First grade	2·5–6·0	50–57	50–56	4·85–5·2

results or trained flavour panels. The usefulness of a non-sensory method for the assessment of cheese quality is usually judged according to how well the non-sensory measurement correlates with sensory data. This creates an interesting situation because if the sensory assessment is carried out by experienced company or state graders this automatically infers that the existing method of grading is acceptable, in which case the purpose for developing non-sensory methods of assessment no longer exists, unless of course one merely wishes to grade a larger proportion of cheese. Although Gilles & Lawrence did not include statistical data relating compositional and grading data, it would not be surprising if the relationship was not particularly good. The difficulty with this situation is determining whether deviations between the two methods are due to the fact that the compositional specifications are not complete or whether the sensory data is inaccurate. It would appear from Gilles and Lawrence's work that both their compositional and their sensory data could be subject to error. Sensory data, because it is entirely subjective, obviously lacks precision and according to Gilles & Lawrence other factors such as the rate and extent of acid production can complicate the use of compositional analyses. Nevertheless despite these obvious difficulties, there is no doubt that compositional analysis can be successfully used provided it is applied to the classification of cheese into a limited number of grades. The introduction of the cheese 'mark' in the UK is an obvious situation to which assessment of quality based upon compositional analysis could be of value, since the aim is only to state that a cheese is above a specified minimum quality. However, it is unlikely that this system of grading would be able to differentiate between the personal preferences of the consumer or the retailer.

4. ASSESSMENT OF QUALITY BY VOLATILE COMPOSITION

There is a fundamental difference between methods for assessing cheese flavour based upon compositional analysis and volatile analysis. The former is based upon the observation that particular ranges in composition tend to yield cheeses of a particular quality standard, whereas the latter is based upon measurements carried out on the volatiles which actually contribute to the flavour. The possibility that the quality of Cheddar cheese could be assessed from measurements of volatile composition was first indicated by Manning (1976). In the course of this work which was aimed at determining which flavour compounds were likely to have the greatest influence on Cheddar flavour, it was found that some compounds, in particular methanethiol, could be related directly to the quality of the cheese as assessed by a trained flavour panel. Further processing of these data has shown that quality can be better defined if the concentrations of other volatiles are included in the relationship. The statistical techniques involved in this type of analysis are considered in more detail later.

The study of volatile composition, normally carried out by head-space analysis combined with gas chromatography, can provide information about both the positive attributes of flavour and, by measuring volatiles known to be responsible for particular defects, about the factors which detract from the quality. Manning (1979) showed, for example, that although measurements of methanethiol composition could provide an assessment of flavour, compounds such as hydrogen sulphide and ethanol, when above a particular concentration, could be related to the defects 'sulphide' and 'fruity' respectively.

These results emphasize the point that some volatile components, for example hydrogen sulphide, although accepted as part of Cheddar flavour at some concentrations, detract from the flavour and become defects when these concentrations are exceeded. Since this is likely to apply to a number of components, one can visualize a situation where the flavour of a cheese will be characterized by specific compounds being present within specific ranges of concentrations.

The association between ethanol concentration and fruity defect in Cheddar is well established, but cheeses with high ethanol concentrations will not necessarily have a fruity flavour. The use of ethanol concentration as an indicator of fruitiness is an example of the use of headspace data to assess the potential quality of the cheese. Fruitiness is a defect which can be fairly common in cheeses which have been stored for a period in excess of 9 months and considerable sums of money could be saved if it were possible to predict whether a young cheese would develop this defect if stored for long periods. It has been shown (Manning, 1979) that if young cheeses contain significantly high concentrations of ethanol then they will probably develop fruity defects during ripening. The reason for this is that fruitiness in cheese arises from the production of ethyl esters during ripening and since Cheddar usually contains an excess of fatty acids the rate of production of the esters is controlled by the availability of ethanol in the cheese. Although observations of ethanol content in young cheeses can provide some insight into the potential of the cheese, for quality control of mature cheeses, analysis of the esters present would provide a better assessment of fruitiness.

It has been mentioned previously that the assessment of the quality of cheese can be made from the concentrations of compounds which contribute to the desirable flavour of the cheese and those which are responsible for defects which detract from the flavour. The relative importance of those two classes of volatiles will depend upon the age of the cheese. For example, since most flavour defects manifest themselves in Cheddar after the cheese has been ripened for at least 7 months, the flavour and quality of cheese younger than this will be largely determined by those volatiles which contribute to its basic and characteristic flavour. On the other hand it has been shown in a recent study carried out at the NIRD that the quality of mature cheeses is largely determined by the extent to which defect flavours develop in the cheese.

The method of volatile analysis for the assessment of cheese flavours, although more complex than that based on gross composition, is likely to provide a more detailed description of the cheese. A large part of this chapter is therefore devoted to volatile analysis of cheese. If volatile analysis is to provide a method of assessing cheese flavour it must be rapid and easily carried out by non-scientific personnel and therefore a detailed account is included on headspace methods which have been found to be suitably applicable to cheese. The emphasis on volatiles is based upon the conclusion that it is the aroma which imparts the characteristic flavour to cheese. McGugan *et al.* (1979) considered that the aqueous fraction of cheese makes the greatest contribution to flavour intensity but they also concluded that it is the volatiles that probably impart the characteristic flavour to Cheddar cheese. However, it would appear from McGugan's studies that it might be possible to obtain some assessment of flavour intensity from the non-volatile fraction of cheese, but this would almost certainly have to be considered together with its volatile composition.

5. REDOX POTENTIAL IN CHEESE RIPENING

Both the methods described for the non-sensory assessment of cheese, if not established, are almost certain to come into use in the future. However, there is another approach which is probably not being seriously considered at this time. It has been concluded by both Law *et al.* (1976) and Manning (1979a) that an essential role of starter in Cheddar cheesemaking is to produce a suitable environment in the cheese for essential volatiles to be produced. During the ripening of Cheddar there is a gradual decrease in redox potential due to the metabolism of the starter or non-starter flora (Kristofferson, 1967). It has been shown that this decrease in redox potential is accompanied by an increase in concentrations of methanethiol and hydrogen sulphide (Green & Manning, 1982), compounds which are accepted as essential to Cheddar flavour. A reducing environment in the cheese is essential for the production and stability of these compounds and it has been shown by Kristofferson and Gould, (1960) that if the redox potential does not decrease sufficiently, these compounds are not formed and the cheese lacks flavour. It is also likely that if fresh surfaces of cheese are exposed to air during cutting the redox potential will rise, the sulphur compounds will become oxidized, with a subsequent fall in flavour quality (Manning *et al.,* 1983). In the light of these findings measurements of redox potential of cheese during ripening may provide valuable information as to flavour development.

6. ANALYSIS OF NEUTRAL VOLATILE COMPOUNDS IN CHEESE

6.1. Isolation of Volatiles

Analysis of the neutral volatile compounds in cheese is invariably carried out using gas chromatography, but before analysis these volatiles have to be isolated from the cheese. Various methods have been used to isolate the volatiles but most suffer drawbacks because of variable recoveries and artifact formation.

Distillation of the lipid fraction (Jackson, 1958; McGugan & How-sam, 1962; Libbey *et al.*, 1963) after separation of the fat by centrifugation produces many volatiles but its adoption assumes that no important volatiles are lost on warming the cheese to 40°C, and that the important volatiles reside in the lipid fraction and not in the aqueous phase. Vacuum distillation of the whole cheese (Liebich *et al.*, 1970a; Manning & Robinson, 1973) avoids the centrifugation step, but is still liable to artifact formation (Lawrence, 1963) unless distillation is carried out at low temperatures. Direct introduction of the cheese oil onto the gas chromatograph column has also been tried (Liebich *et al.*, 1970b).

Solvent extraction of the whole cheese (Liebich *et al.*, 1970a) has also been used to isolate the volatiles but poor yields of the water soluble materials are inevitable. The solvents used must be of high purity to prevent artifact formation and concentration of the extract leads to losses of the more volatile compounds.

The methods so far described have been found suitable for qualitative analysis of cheese volatiles but because of variable recoveries and poor reprodubility they are unsuitable for the quantitative analysis necessary for the assessment of cheese quality. These methods are also too time-consuming to be suitable for routine analysis.

A preferred method, headspace analysis, utilizes the equilibrium that exists between the volatile components of a liquid or solid sample and the surrounding gas phase in a sealed vessel. In the case of cheese the headspace is the vapour mixture emanating from the surface of the cheese. Using headspace analysis quantitative results can be achieved provided the actual concentration of the components within the cheese is not required. Cheese is a complex system consisting of lipids, protein, carbohydrates, salts and water, and to relate the concentrations of volatiles in the headspace to their concentrations in the cheese, it would be necessary to know how the volatiles are partitioned between the various components.

The headspace over cheese can be analysed by introducing it directly into a gas chromatograph. Alternatively an extended headspace technique can be used where larger volumes of the headspace are concentrated prior to introduction into the gas chromatograph. When applying a headspace technique to the analysis of solids it is necessary to increase the surface area by reducing the particle size to allow the volatiles present in the solid to quickly equilibrate with the surrounding headspace. Bottazzi & Battistotti (1974) when studying different varieties of Italian cheese, grated the cheese into a flask which could be closed with a rubber seal. But as low concentrations of very volatile sulphur compounds are important to cheese flavour, notably with Cheddar cheese (Manning & Robinson, 1973; Manning, 1974; Manning & Price, 1977), these compounds can easily be lost when the cheese is grated, being adsorbed onto the surface of the vessel or absorbed by the rubber closure (Gilliver & Nursten, 1972).

The method developed by Manning *et al.* (1976) has overcome these drawbacks by using the cheese itself as the container for the headspace. The cheese is allowed to warm in a temperature-controlled room to 20°C prior to analysis. A core is removed from the sample using a standard cheese borer and the entrance of the bore hole immediately closed with a stainless steel cap fitted with a septum as shown in Fig. 1. After a period of 1 h to allow the volatiles within the bore hole to come to equilibrium with the cheese, samples of the headspace are removed via the septum using a gas syringe, and transferred to the gas chromatograph. The volume of the bore is approximately 30 ml and up to 7 ml can be removed before there are significant losses of volatiles. Figure 2 shows the headspace chromatograms of a series of commercial hard cheeses produced using this method (Manning & Moore, 1979).

Manning's method for sampling the headspace of hard cheese has proved to be both reliable and reproducible, but suffers from the disadvantage that the cheese must be allowed to warm up to about 20°C before sampling. This means that even for a small cheese (4·5 kg) it is necessary to leave them in a temperature-controlled room overnight to warm from store to ambient temperature. This is not practical for the routine sampling of cheese and Price & Manning (1983) have developed a technique which overcomes this problem and enables the cheese to be sampled at cold store temperatures (6°C). Their device for sampling the cheese and producing the headspace is shown in Fig. 3. It consists of two parts: a sampler assembly and an extrusion assembly. The sampler assembly consists of a borer through which a plunger operates. It is pushed into the cheese which can be at cold store temperature and then withdrawn so that a cylinder of cheese (20 mm diameter × 65 mm) is retained within the borer. The open end is then closed with a cap and the borer is placed in a water bath at 30°C for 15 min to allow the cheese to equilibrate to that temperature. The extrusion assembly is made up of 4 parts: a stainless steel barrel and extrusion plate, a polytetrafluoroethylene (PTFE) main chamber and end plate with a septum port, which bolt together to form a gas-tight

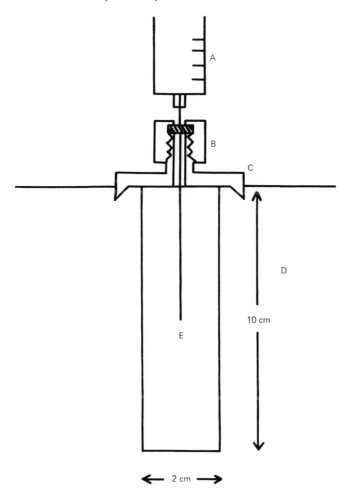

Fig. 1. Device used to obtain headspace samples from cheese for gas chromatography. A, gas syringe; B, septum holder; C, stainless steel cap; D, cheese; E, bore hole (from Manning *et al.* (1976), reproduced with the kind permission of *J. Dairy Res.*).

chamber with the extrusion plate positioned between the chamber and the barrel. When the cheese sample has reached equilibrium temperature the sample assembly is screwed into the barrel of the extrusion assembly. By pressing the plunger the cheese is forced through the extrusion plate and into the main chamber which increases the surface

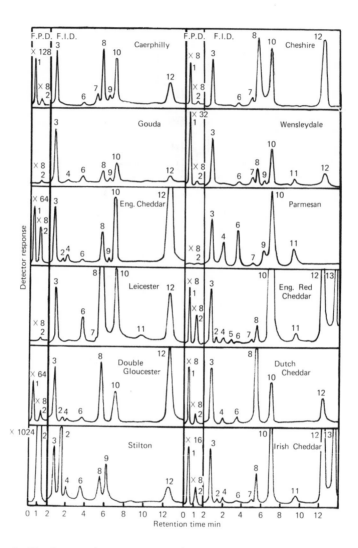

Fig. 2. Headspace chromatograms of commercial hard cheeses. Peak identification: 1, hydrogen sulphide; 2, methanethiol; 3, air peak + diethyl ether; 4, acetaldehyde; 5, propanal; 6, acetone; 7, ethyl acetate; 8, butanone; 9, methanol; 10, ethanol; 11, pentan-2-one; 12, butan-2-ol; 13, propan-1-ol (from Manning & Moore (1979), reproduced with the kind permission of *J. Dairy Res.*).

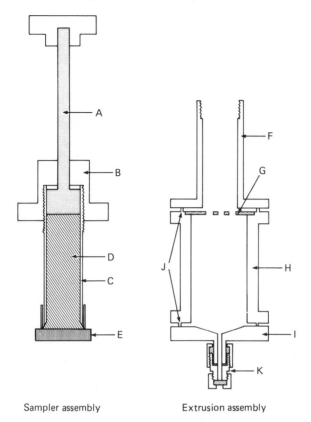

Sampler assembly   Extrusion assembly

Fig. 3. Device used to obtain headspace samples from cheese for gas chromatographic analysis. A, plunger; B, handle; C, borer; D, cheese core; E, cap; F, barrel; G, extrusion plate; H, teflon chamber; I, end plate; J, sealing lips; K, septum port (from Price & Manning (1983), reproduced with the kind permission of *J. Dairy Res.*).

area of the cheese and allows the volatile substances present in the cheese to quickly come to equilibrium with the surrounding headspace. The optimum time for equilibrium to be established is 5 min after which samples of the headspace can be withdrawn via the septum port with a gas-tight syringe and transferred to the gas chromatograph.

Similar results are obtained using the cavity method but the extrusion system has the advantage that sampling of the cheese can be done at ripening temperatures. Also, since cheese samples stored within the

TABLE 2
Concentration of Volatiles Typically Found in the Headspace over Mature
Cheddar Cheese by Direct Headspace Method

	pg/ml		*ng/ml*		*ng/ml*
Hydrogen sulphide	2000	Acetone	10	Ethanol	30
Methanethiol	800	Butanone	10	Butan-2-ol	1
Dimethyl sulphide	200	Pentan-2-one	1	Ethyl acetate	0·5

capped sampling assembly can be kept at 4°C for up to 3 h before extrusion without affecting the concentration of volatile compounds, the method is suitable for routine sampling of cheese in a factory. Typical levels of volatiles found in Cheddar cheese using this method are shown in Table 2.

The direct headspace methods described are quick to use, do not require scientific personnel to operate and can therefore easily be adapted for routine quality control in the creamery. They do however suffer from the disadvantage that they are biased towards the most volatile components, those having appreciable vapour pressures. This means that it is possible that some high boiling, low threshold flavour compounds could go undetected. It is this factor which has led to the development of extended headspace methods.

The extended headspace technique has in recent years been applied to the analysis of many foods and environmental pollutants (Kolb, 1980), and Horwood (1974) has applied the technique to Cheddar and Cheedam cheese. Shredded cheese was placed in a flask previously flushed with helium and after allowing the cheese to stand for at least 16 h a headspace trap consisting of a glass tube containing Porapak Q was connected to the flask. Helium was passed through the sample and out through the Porapak trap where the volatiles were concentrated and water vapour allowed to pass straight through. The trap containing the volatiles was connected to the front of a gas chromatography column and the volatiles desorbed by heating the trap to 195°C for 30 min. Horwood's method produced 9 GLC peaks from Cheddar cheese, the main volatiles being diacetyl, pentanal, pentan-2-one, and heptan-2-one. Horwood's results confirmed that recoveries were better for the less volatile compounds but many very volatile compounds, found using the direct headspace method, were lost.

Regeneration of trapped volatiles from porous polymers (Porapak or Tenax) by heating tends to produce aromatic hydrocarbons, toluene, xylenes and ethylbenzene due to breakdown of the polymer. However the availability of Tenax-TA, a specially processed Tenax designed for use as a trapping agent and having very low levels of these potentially interfering impurities, should minimize artifact formation. Activated charcoal has also been successfully employed as a trapping agent for headspace volatiles (Clark & Cronin, 1975; Grob & Zurcher, 1976), regeneration of the volatiles being carried out either thermally or chemically. With thermal desorption, high temperatures are required (>260°C) which could lead to artifact formation, whereas chemical desorption with a high affinity solvent such as carbon disulphide can lead to the solvent peak masking important areas of the chromatogram. It also requires that the solvent should be of high purity.

Cryogenic trapping of volatiles eliminates problems associated with regeneration, but the high moisture content of cheese results in large amounts of water being co-trapped, with the volatiles; this leads to traps becoming blocked with ice and excess amounts of water being transferred to the chromatograph.

6.2. Gas Chromatographic Analysis

Analysis of volatiles isolated from cheese can be carried out using either conventional packed columns or glass capillary columns. Conventional packed columns have tended to be preferred, because up to 5 ml of vapour can be directly injected onto the column without loss of resolution. Also, bearing in mind that this type of analysis is intended to be carried out in factories by non-scientific personnel, they are much simpler to use. However, these columns do not have high separating powers and since a wide range of classes of compounds are usually encountered in flavour analysis, a single column is often insufficient to adequately separate a wide range of compounds. Manning (1978b), for example, used two chromatographs for his analyses. One was fitted with a 1·5 m × 4 mm i.d. glass column packed with 15% Carbowax 20 M on Universal B support and equipped with a flame ionization detector (FID) for the analysis of non-sulphur volatiles. The other was equipped with a flame photometric detector (FPD), washing in the sulphur mode, for the sulphur volatiles.

The FPD is not only specific for sulphur compounds but for these compounds it has a much greater sensitivity than the FID. As the sulphur compounds were present at very low concentrations and were quite reactive, Carbopack B-HT-100, a hydrogen treated, deactivated, graphitized carbon absorbent (Supelco Inc., Bellefonte, Pa, USA), specifically developed for the analysis of sulphur compounds, was used (Manning, 1978b). More recently Price & Manning (1983) have modified the chromatographic conditions to enable the entire analysis to be carried on a single column. They used a single column containing Carbopack B-HT-100 for the separation of both sulphur and nonsulphur containing volatiles but the eluent from the column was split (1:1) to take the sample to a FID and FPD simultaneously.

The advantages of packed columns have already been mentioned, but they suffer from the disadvantage that having a relatively low number of theoretical plates, they have low resolving power. Wallcoated open tubular (WCOT) columns on the other hand have high separating efficiency, thermal stability and few active sites, resulting in good peak shapes for even the most polar compounds. WCOT columns made of glass or silica are now commercially available and produce narrow peak widths providing low limits of detection. However, they do have very low loading capacities (20 ng/component) and the stationary phases are easily damaged. The recent introduction of wide bore (0·3 mm) thick film bonded phase silica columns has improved the use of capillary columns on a routine basis, because they are virtually unbreakable and have an almost indestructable stationary phase.

Capillary columns are ideally suited to analysis of volatiles by extended headspace techniques. If volatiles are regenerated from traps by thermal desorption they can be swept directly onto the column and retrapped in a narrow zone by cooling the first few centimetres of the column. Removal of the coolant and initiation of oven temperature programme starts the separating process. Alternatively if volatiles are eluted from charcoal with solvent, up to 2 μl of the extract can be directly introduced onto the column.

7. SENSORY ASSESSMENT OF CHEESE FLAVOUR

7.1. Grading

It has previously been said that in order to develop non-sensory methods for assessing cheese flavour, the results must be related to sensory data. This may take the form of graders' reports or the assessment of the cheese by a trained flavour panel. Either method is acceptable providing it is understood that the type of sensory data obtained from the two approaches will be somewhat different.

Company graders have invariably classified cheeses within up to five grades ranging from 'extra selected' to 'no grade' and this has formed the basis of the New Zealand and Australian approach to developing non-sensory assessment methods. This approach has the advantage that limits for compositional or volatile analyses can be set up for different grades of cheeses. The disadvantage of the approach is that cheeses will be graded not only according to their flavour and aroma but also according to body and texture, colour and possibly finish. It is therefore less likely that good correlations will be obtained between some chemical or volatile parameters and a summation of a number of aspects of flavour. An improvement in classifying cheeses into grades, would be the use of the grading scheme used in the past in the UK by the National Association of Creamery Proprietors and Wholesale Dairymen. This scheme involved awarding marks out of 45 for flavour and aroma, marks out of 40 for body and texture, marks out of 10 for colour and marks out of 5 for finish, giving a total mark out of 100 for the cheese. Using this approach it becomes possible to correlate a particular non-sensory parameter with the particular aspect of flavour to which it is more applicable.

7.2. Panels

The approach adopted in the UK has been to use trained flavour panels to evaluate quality based upon flavour and aroma. The panels, which consist of about 24 tasters, are required to assess the quality of the cheese on a scale of 0 (very poor) to 8 (excellent) and the intensity of flavour also on a scale of 0 (absent) to 8 (very strong). They are also asked to assess flavour defects on a scale of 0 (absent) to 4 (very strong). In assessing the quality of the cheese, panellists are looking for those flavour notes which go to make up the characteristic flavour of the cheese and those which are clearly defects and detract from the quality. A cheese can therefore attain a poor quality score either because it lacks the necessary flavour notes or because some components are present at too high a concentration and impart a flavour defect. The advantage of this method is that one can seek correlations between non-sensory parameters and either the positive attributes of the cheese or specific flavour defects. Whether the sensory data are obtained from commercial graders or from trained panels they are still amenable to the statistical techniques outlined in the next section.

8. STATISTICAL METHODOLOGY IN THE NON-SENSORY ASSESSMENT OF FLAVOUR

Persson *et al.* (1973) state that 'In order to be able to interpret sensory data and their changes, it is essential to analyse the material instrumentally for chemical properties and to develop models for correlating the sensory and instrumental data in a meaningful way.' Such experiments frequently generate large amounts of data (e.g. data from complex gas chromatograms and flavour panels involving several trained judges) and hence necessitate the use of computers if thorough and accurate analyses are to be completed within practical time limits.

Many varied statistical techniques have been applied to such data, though the most common preliminary approach has been that of simple parametric analyses, such as the analysis of variance, correlations and regressions. These techniques have been applied to a wide range of food products (Kristofferson & Gould, 1960; Powers, 1968; Teranishi *et al.*, 1971; Dravnieks *et al.*, 1973; Harries *et al.*, 1974; Manning *et al.*, 1976; Khayat, 1979; Lawrence & Gilles, 1980).

The analysis of variance and linear regression are known to be reliable statistical techniques. However, they do depend upon various underlying assumptions which must be fulfilled if the inference resulting from their application is to be relied upon. The algebraic form of the simple linear regression is as follows:

$$y_i = a + bx_i + e_i$$

where the subscript i refers to the ith observation, y_i is the dependent variable, x_i is the independent variable, a is the intercept (the value of y at $x = 0$), b is the gradient or slope of the line and e_i is a residual term which allows for the fact that observed values y_i are likely to differ from the exact linear relationship $a + bx_i$. The residual e_i is often referred to as an error term. It is assumed that the residuals are normally distributed and have the same variance. They should also be statistically independent, i.e. the probability that the error of any observation has a particular value must not depend upon the values of the errors for other observations. Cochran (1947) concludes that 'In general, the factors that are liable to cause the most severe disturbances are extreme skewness, the presence of gross errors, anomalous behaviour of certain treatments or parts of the experiment, marked departures from the additive relationship and changes in the error variance, either related to the mean or to certain treatments or parts of the experiment. The principal methods for an improved analysis are the omission of certain observations, treatments or replicates, subdivision of the error variance, and transformation to another scale before analysis.' However, linear regression and the analysis of variance are regarded as 'robust' statistical techniques, and small departures from the outlined assumptions can be tolerated without marked effects on the overall conclusions.

It has been stressed that the final standard of quality is human evaluation and hence objective procedures (i.e. non-sensory analyses) should be judged by their correlation with sensory results (Amerine *et al.*, 1965).

The correlation coefficient, *r*, measures the degree of linear association between data sets. Some research workers even consider that coefficients of correlation must reach a fixed minimum value in order to be in any way useful (Kramer & Twigg, 1970). Some correlations between the headspace volatiles of Cheddar cheese and sensory data have been calculated. Kristofferson & Gould (1960) found a positive correlation between hydrogen sulphide concentration and flavour scores, and Manning *et al.* (1976) found a positive correlation between the intensity of Cheddar flavour and methanethiol concentrations in cheese headspace. Research in New Zealand has shown that 4 physicochemical factors largely determine the quality of Cheddar cheese, though no formal statistical analyses are reported (Gilles & Lawrence, 1973; Lawrence & Gilles, 1980). In some cases, therefore, simple correlation methods may suffice because one component is closely correlated with flavour quality or perhaps a particular off-flavour of the food product.

Regression coefficients provide an estimate of the best straight line that can be fitted through the data points and therefore reveal more detail about the relationship than correlation coefficients alone. Often the relationship between sensory and non-sensory variates is not linear, and transformations are necessary. The most frequently attempted transformation is to logarithms (Dravnieks *et al.*, 1973; Khayat, 1979). Von Sydow (1971) states that it is generally recognized that response values and stimulus values follow Stevens' Law:

$$R = C . S^n$$

which he has used as a model for relating instrumental (*S*) and sensory (*R*) data in its power function as above, but also in its linear logarithmic form,

$$\log R = n \log S + \log C$$

where n and C are constants.

A natural extension of such regression analyses is the powerful technique of multiple regression, and in particular, stepwise multiple regression where several independent variates are used to describe the variation in the dependent variate. These methods have been used widely in analysing the relationships between sensory and non-sensory data (Dravnieks *et al.*, 1973; Kosaric *et al.*, 1973; Galetto & Bednarczyk, 1975; Manning, 1976; Khayat, 1979; Manning, 1979). Flavour differences are often subtle and can arise from minor differences in several components rather than a striking change in one compound. Kramer & Twigg (1970) describe the application of stepwise multiple regression to instrumental and sensory data. The taste panel score is considered as the 'dependent' or 'y' variable, and the 'independent' or 'x' variables are provided by individual non-sensory components such as the headspace volatiles of Cheddar cheese which may contribute to the flavour sensation. Data for each component are correlated with each other, to form a correlation matrix. The highest correlation coefficient between panel results and one of the independent variables is selected, and all of the other correlations are recalculated on a partial correlation basis. The next highest correlation coefficient is then selected, and if the multiple correlation coefficient is significantly better than the previous single high correlation, both are retained for use in the regression equation, and the remaining correlations are recalculated, omitting the two selected. This process is repeated until all of the components which contribute to a significant increase in the multiple correlation coefficient with the taste panel data have been selected. In this way, the components having a significant effect on flavour perception are selected while the other components are omitted from the final regression equation. The regressions are thus essentially linear. This technique has been successfully used to relate contributions of individual flavour components to the flavour of Cheddar cheese by Manning (1979). Equations were derived which related Cheddar quality (CQ) to the headspace concentrations in ng/5 ml of 2-pentanone (C_3), methanethiol (C_1) and methanol (C_2) as follows:

$$CQ = 0 \cdot 25C_1 + 0 \cdot 08C_2 + 0 \cdot 018C_3 + 2 \cdot 04.$$

These 3 volatiles accounted for 59% of the panel's response for

Cheddar quality. Similarly the derived multiple regression equation relating Cheddar intensity (CI) to concentrations of volatiles was:

$$CI = 0.35C_1 + 0.017C_2 + 0.07C_3 + 0.47$$

accounting for 71% of the panel's response to Cheddar intensity. Care must be taken in the interpretation of the individual components of multiple regressions. Each coefficient is calculated on a partial correlation basis which takes into account correlations between the independent variables themselves. The value of a component coefficient in a multiple regression will not be the same as its value when fitted alone.

Gas chromatograms have been commonly used as measures of non-sensory attributes of sensory compounds (Powers, 1968; Powers & Keith, 1968; Dravnieks *et al.*, 1973; Persson & Von Sydow, 1973; Persson *et al.*, 1973; Gostecnik & Zlatkis, 1975; Manning, 1976; Manning, 1979; Khayat, 1979; Mottram *et al.*, 1982). The methods of analysis so far discussed require that the identity of the chemical components detected by the gas chromatogram be established. These methods do not involve the use of the entire chromatogram, but rather one or several peaks among all those detected. This is satisfactory if a good correlation exists between flavour and a few chemical compounds.

Another possible approach is to use the gas chromatographic patterns directly. Each sample is considered as an assembly of variables, with each peak being an individual variable. Foods judged different in flavour may often yield chromatograms whose patterns differ, but the change in any one peak, or in relatively few peaks, is so minor that correlation with sensory differences is poor. Powers (1968) states that 'Flavour is usually a consortium among compounds. The ratios of the various substances one to another determine the dominant flavour and the various flavour notes.' Such data require analysis by multivariate methods because of their complexity, and a statistical technique which lends itself well to the analysis of gas chromatograms is that of discriminant analysis. Discriminant analysis was developed by Fisher (1936) who sought a mathematical means of deciding to which of two varieties unknown specimens of iris belonged. The varieties differed in several characters, but for each characteristic there was so much overlap that one variable alone was insufficient to distinguish one variety from another. When all the variables were considered, a decision could be made on the cumulative evidence with rather less doubt. The peaks on the chromatograms can be considered to be comparable to the characteristics of the iris varieties. In some cases, no one peak is sufficient to distinguish one sample from another.

Powers & Keith (1968) found that they could carry out a stepwise discriminant analysis better if the ratios between the peak heights were used instead of the peak heights themselves, though he felt that peak area should be selected in most instances (Powers, 1968). Dravnieks *et al.* (1973) suggest transformations on the peak area data including various ratios, percentages and logarithms.

The function of discriminant analysis is to find which combinations of variables (e.g. gas chromatography peaks) are most useful in discriminating between the categories of the food samples. It can therefore be regarded as a procedure of classification. In the stepwise discriminating program, this is achieved by taking first the most discriminating peak, then selecting the next most discriminating etc., until the useful information is exhausted. The technique has been used successfully in the classification of flavour by several research workers (Powers, 1968; Powers & Keith, 1968; Dravnieks *et al.*, 1973; Gostecnik & Zlatkis, 1975). Further discriminating techniques include canonical variate analysis which has been used successfully (coupled with stepwise discriminant analysis) to unambiguously distinguish lean meats and lean meats cooked with adipose tissue. Variables were obtained by gas chromatography–mass spectrometry and panel evaluation of meat odour (Mottram *et al.*, 1982). Canonical variate analysis has also been applied to data obtained from gas–liquid chromatography techniques to discriminate between genus groups of food spoilage bacteria (MacFie *et al.*, 1978; O'Donnell *et al.*, 1980).

These and several other methods of multivariate data analysis offer ways of simplifying and interpreting many different variables simultaneously. They can reveal the main structures and relationships in large data tables, giving relatively simple output graphs and tables which have a maximum of information and a minimum of repetition. Martens & Russwurm (1983) claim that multivariate methods 'may yield conclusions that are more precise than the individual input data because by analysing many variables from many different objects simultaneously, the bad effects of random noise in the input data is reduced, much like averaging over replicate measurements.' They also feel that multivariate methods can compensate for many types of systematic errors in the input data and that they are far superior to traditional univariate methods with respect to outlier detection. A comprehensive series of useful papers illustrating many of the multi-variate techniques available for the analysis of both sensory and non-sensory food data can be found in Martens & Russwurm (1983). The computations for the many methods of analysis of instrumental and sensory data, whether multivariate or univariate, are efficiently carried out with the aid of a statistical programming language such as GENSTAT (GENSTAT Manual, 1977).

Whatever the method of analysis, one of the most important exercises is to look at graphs of the data. This applies for simple raw data plots as well as plots from the output of complex multivariate programs. Graphs are a very good analytical tool which can provide rapid interpretation of the main trends and patterns within the data, and their usefulness should not be underestimated.

9. CONCLUSIONS

In this chapter we have tried to emphasize the need for non-sensory methods for assessing cheese flavour. Examples have been given of how, individually, chemical and headspace can each provide information about the quality and potential quality of cheese, but in a commercial situation data obtained by a combination of the different approaches will be needed. Methods used in chemical analysis have not been included as these are well established, but headspace techniques using the latest developments on gas chromatography have been discussed.

Development of non-sensory methods for cheese necessarily requires that non-sensory data should be related to sensory data obtained from either panels or graders, and some discussion has been included on the techniques used in these two approaches. In the initial stages panels are probably the most useful as they can be trained to assess any particular aspect of flavour, but if the method is to be used commercially then the non-sensory data have ultimately to be correlated with graders' assessments.

Statistical methodology forms an essential part in relating non-sensory to sensory data and for this reason a large part of this chapter has been devoted to this subject. The entire field of non-sensory assessment of food flavours is expanding, but it is the area of statistical methods in which we are likely to see the greatest developments in the future.

REFERENCES

AMERINE, M. A., PANGBORN, R. M. & ROESSLER, E. B. (1965) *Principles of Sensory Evaluation of Food.* Academic Press, New York.

BOTTAZZI, V. & BATTISTOTTI, B. (1974) *Scienza e Tecnica Lattiero–Casearia* **25,** 11.

CLARK, R. G. & CRONIN, D. A. (1975) *J. Sci. Food Agric.* **26,** 1615.

COCHRAN, W. G. (1947) *Biometrics* **3**(1), 22.

DRAVNIEKS, A., REILICH, H. G. & WHITFIELD, J. (1973) *J. Food Sci.* **38**(1), 34.

FISHER, R. A. (1936) *Ann. Eugenics (London)* **7**(2), 179.

GALETTO, W. G. & BEDNARCZYK, A. A. (1975) *J. Food Sci.* **40**(6), 1165.

GENSTAT MANUAL (1977) GENSTAT, a general statistical program, Numerical Algorithms Group, Oxford.

GILLES, J. & LAWRENCE, R. C. (1973) *New Zealand J. Dairy Sci. Technol.* **8,** 148.

GILLIVER, P. J. & NURSTEN, H. E. (1972) *Chem. & Ind.* **13,** 541.

GOSTECNIK, G. F. & ZLATKIS, A. (1975) *J. Chromatog.* **106,** 73.

GREEN, M. L. & MANNING, D. J. (1982) *J. Dairy Res.* **49,** 737.

GROB, K. & ZURCHER, F. (1976) *J. Chromatog.* **117**(2), 294.

HARRIES, J. M., POMEROY, R. W. & WILLIAMS, D. R. (1974) *J. Agric. Sci. Cambridge* **83,** 203.

HORWOOD, J. F. (1974) *19th Int. Dairy Congress* **1E,** 503.

JACKSON, H. W. (1958) *Perfumery & Essential Oil Record* **49,** 256.

KHAYAT, A. (1979) *J. Food Sci.* **44**(1), 37.

KOLB, B. (Ed.) (1980) *Applied Headspace Gas Chromatography.* Heyden, London.

KOSARIC, N., DUONG, T. B., SVRCEK, W. Y. (1973) *J. Food Sci.* **38**(3), 369.

KRAMER, A. & TWIGG, B. A. (1970) *Quality Control for the Food Industry,* 3rd edn, Vol. 1. AVI Publishing Company Incorporated, Westport, Connecticut.

KRISTOFFERSON, T. (1967) *J. Dairy Sci.* **50,** 279.

KRISTOFFERSON, T. & GOULD, I. A. (1960) *J. Dairy Sci.* **43,** 1202.

LAW, B. A., CASTANON, M. J. & SHARPE, M. E. (1976) *J. Dairy Res.* **43,** 301.

LAWRENCE, R. C. (1963) *J. Dairy Res.* **30,** 161.

LAWRENCE, R. C. & GILLES, J. (1980) *New Zealand J. Dairy Sci. Technol.* **15,** 1.

LIBBEY, L. M., BILLS, D. D. & DAY, E. A. (1963) *J. Food Sci.* **28,** 239.

LIEBICH, H. M., DOUGLAS, D. R., BAYER, E. & ZLATKISS, A. (1970a) *J. Chromatog. Sci.* **8,** 355.

LIEBICH, H. M., DOUGLAS, D. R., BAYER, E. & ZLATKISS, A. (1970b) *J. Chromatog. Sci.* **8,** 351.

McBRIDE, R. L. & HALL, C. (1979) *The Australian J. Dairy Technol.* June, 66.

MACFIE, H. J. H., GUTTERIDGE, C. S. & NORRIS, J. R. (1978) *J. Genetics Microbiol.* **104,** 67.

McGUGAN, W. A. & HOWSAM, S. G. (1962) *J. Dairy Sci.* **45,** 495.

McGUGAN, W. A., EMMONS, D. B. & LARMOND, E. (1979) *J. Dairy Sci.* **62,** 398.

MANNING, D. J. (1974) *J. Dairy Res.* **41,** 81.

MANNING, D. J. (1976) PhD Thesis, Reading University/Physics Department, NIRD.

MANNING, D. J. (1978a) *Dairy Industries Int.* **43**(4), 37.

MANNING, D. J. (1978b) *J. Dairy Res.* **45,** 479.

MANNING, D. J. (1979) *J. Dairy Res.* **46,** 523.

MANNING, D. J. & MOORE, C. (1979) *J. Dairy Res.* **46,** 539.

MANNING, D. J. & PRICE, J. C. (1977) *J. Dairy Res.* **44,** 357.

MANNING, D. J. & ROBINSON, H. M. (1973) *J. Dairy Res.* **40,** 63.

MANNING, D. J., CHAPMAN, H. R. & HOSKING, Z. D. (1976) *J. Dairy Res.* **43,** 313.

MANNING, D. J., RIDOUT, E. A., PRICE, J. C. & GREGORY, R. J. (1983) *J. Dairy Res.* **50,** 527.

MARTENS, H. & RUSSWURM, H. JR. (1983) *Food Research and Data Analysis.* Applied Science Publishers, London & New York.

MOTTRAM, D. S., EDWARDS, R. A. & MACFIE, H. J. H. (1982) *J. Sci. Food Agric.* **33,** 934.

O'DONNELL, A. G., MACFIE, H. J. H. & NORRIS, J. R. (1980) *J. General Microbiol.* **119,** 189.

PERSSON, T. & VON SYDOW, E. (1973) *J. Food Sci.* **38**(3), 377.

PERSSON, T., VON SYDOW, E. & AKESSON, C. (1973) *J. Food Sci.* **38**(4), 682.

POWERS, J. J. (1968) *Food Technol.* **22**(4), 39.

POWERS, J. J. & KEITH, E. S. (1968) *J. Food Sci.* **33,** 207.

PRICE, J. C. & MANNING, D. J. (1983) *J. Dairy Res.* **50,** 381.

TERANISHI, R., HORNSTEIN, I., ISSENBERG, P. & WICK, E. L. (1971) *Flavour Research–Principles and Techniques.* Marcel Dekker, Inc., New York.

VON SYDOW, E. (1971) *Food Technol.* **25,** 40.

Index